THE ELEMENTS OF EN(

The Elements of Engineering Contracts

M. O'C. HORGAN, OBE, TD, DL,
MSc, CEng, FIEE

and

F. R. ROULSTON, DLC, CEng, MIEE

W. S. Atkins & Partners
Project Management Services Division
Woodcote Grove
Epsom, Surrey, England
(Edition: August 1977)

Whilst every effort has been made to achieve accuracy in the compilation of this book, the publishers accept no liability whatsoever in respect of any errors in or omissions from the contents.

Published by E. & F.N. Spon Limited
11 New Fetter Lane, London EC4P 4EE

©1977 W.S. Atkins & Partners

Printed in Great Britain by
Woodcote Publications Limited, Epsom, Surrey

ISBN 0 419 11610 9

This paperback edition is sold subject to the condition that it shall not, by way of trade or otherwise, be lent, re-sold, hired out, or otherwise circulated without the publisher's prior consent in any form of binding or cover other than that in which it is published and without a similar condition including this condition being imposed on the subsequent purchaser

All rights reserved. No part of this book may be reprinted, or reproduced or utilized in any form or by any electronic, mechanical or other means, now known or hereafter invented, including photocopying and recording, or in any information storage or retrieval system, without permission in writing from the publisher

Preface

This work has been produced in the normal course of their business by W. S. Atkins & Partners primarily to help their own staff in dealing with contractual matters. It forms the first of three volumes, the other two (dealing with Tender Procedure and Contract Control) being essentially of domestic interest. This one, being more general in its outlook, is being made available by W. S. Atkins & Partners to others who may be interested in the contractual aspects of engineering projects.

The authors wish to acknowledge gratefully the help received from Mr. R. K. Corrie, Chief Engineer, Project Management Services Division of W. S. Atkins & Partners (at whose instigation the compilation of this volume was undertaken) more especially for his advice and criticism as the work proceeded.

Thanks are also due to senior colleagues for their kind reception of the first edition of this volume and the suggestions made from their wide experience as to how it might usefully be modified or expanded. Most of their suggestions are now incorporated in this revised edition.

M.O'C.H.
F.R.R.

Contents

Page

Introduction 11

Definitions 13

Chapter 1—What makes a Legal Contract? 17

1.1 An agreement of views between the parties
1.2 Essential elements
1.3 Intention to create a Legal Relationship
1.4 A Genuine Consent of the Parties
1.5 Legal Capacity of Parties to act
1.6 Legality of Objects of the agreement
1.7 Valuable Consideration
1.8 Contracts Verbal and Written
1.9 Breach of Contract

Chapter 2—What does an Engineering Contract usually consist of? 26

2.1 Object
2.2 Contract Documents
2.3 Statement of the scope of the Contract
2.4 Data affecting the execution of the Works
2.5 The Technical Specification
2.6 Bills of Quantities, Prime Cost Items and Provisional Sums
2.7 Programme for the construction/completion of the Works
2.8 The Conditions of Contract
2.9 Site Regulations
2.10 The Tender
2.11 Bonds and Guarantees
2.12 A form of Agreement
2.13 Agreements 'under hand' and 'under seal'
2.14 Subsequent variations to a Contract

Chapter 3—Negotiations leading to a Contract   50

3.1   Need for Negotiation
3.2   Enquiries against Firm Specifications
3.3   Enquiries regarding large and complex projects
3.4   Final Appraisal and Selection

Chapter 4—Awarding the Contract   55

Chapter 5—Notifying the Selected Contractor   57

5.1   Letter of Intent
5.2   Letter of Acceptance
5.3   Instruction to Proceed

Chapter 6—Notes on Standard Forms of Conditions of Contract   60

6.1   Standard Forms of Conditions of Contract in common use
6.2   Standard Forms of Conditions of Contract used by large organisations
6.3   Institution of Civil Engineers
6.4   IMechE/IEE Model Conditions
6.5   Institution of Chemical Engineers
6.6   Standard Form of Building Contract ("RIBA" or "JCT")
6.7   British Electrical and Allied Manufacturers Assoc. (BEAMA)
6.8   Standard Conditions of Purchase of Goods
6.9   International use—FIDIC
6.10   International use—U.N. Economic Commission for Europe
6.11   Requirements of Credit Funding and Insuring Organisations

Chapter 7—Types of Contract   70

7.1   Classified by method of evaluating Contract Price:

     7.1.1   Fixed Price Contract
     7.1.2   Price Adjustment Contract
     7.1.3   Cost-plus Contract
     7.1.4   Target-cost Contract

      7.1.5 Bills of Quantity Contract
      7.1.6 Schedule of Rates Contract

7.2   Classified other than by method of evaluating Contract Price:

      7.2.1 Competitive Contract
      7.2.2 Negotiated Contract
      7.2.3 Package Contract—Package Deal
      7.2.4 Turnkey Contract
      7.2.5 Continuation Contract
      7.2.6 Serial Contract
      7.2.7 Running Contract
      7.2.8 Service Contract

Chapter 8—Assignment and sub-letting of Contracts   84

8.1   Assignment
8.2   Sub-Letting and Sub-Contracting
8.3   Nominated Sub-Contractors

Chapter 9—Types of Tender   90

9.1   Open Tender
9.2   Selective Tender
9.3   Negotiated Tender
9.4   Serial Tender
9.5   Requirements of the World Bank and I.D.A.

Chapter 10—Conditions of Contract printed on order forms   94

Appendix 1   97

Equivalent terms used in some common Standard
Forms of Conditions of Contract

Index   99

Introduction

The object of this Volume is to set out the main features of industrial contracts such as are typically used for buying and selling, for constructing, installing, erecting and similar activities in the fields of civil engineering, plant and machinery, and building works. It is not primarily intended for contract specialists (though they may find it useful for reference) and consequently it makes no attempt to deal with the more legal aspects and complexities of contract conditions: rather it is written with the aim of being readable by those who are either newly concerned with the business of handling engineering contracts or whose experience hitherto has been confined to their more technical aspects. It has been accepted that, as a result, the contents may occasionally not achieve full legal precision and completeness, but it is hoped they will never mislead the reader.

A contract is a formal agreement, enforceable at law, between two (or more) parties whereby each accepts certain responsibilities and receives certain benefits. Failure in any way to carry out a responsibility so undertaken usually involves some form of sanction, either expressed or implied in the contract, or under the law of the land. It is therefore necessary that all people concerned in executing a contract are fully aware of just what the exact responsibilities of the parties to that contract are and the precise implications of failing to fulfil them. The liabilities for breach of contract can be heavy and drastic, easily exceeding the value of the contract itself if proper safeguards are not arranged. Those responsible for implementing a contract must always "watch their step".

The foregoing applies with especial force in the case of representatives of specialist or consultancy firms on whom their Clients rely for sound advice and faultless action in the management and administration of their contracts. At stake such firms have both their reputation for reliability and their freedom from actions for damages for failing to provide due expert knowledge and care in their Clients' interests. Those who sell specialist advice just cannot afford to make mistakes.

Definitions

In this volume we shall standardise the use of the undermentioned words to have the meanings shown. The same will be found in some contracts but in others alternative words having more or less identical meanings are used, for example Purchaser, Buyer, Owner, Promoter will be found instead of Employer; Plant and Equipment will be found reversed and so on. For ease of reference a comparative table of definitions in the more usual Standard Forms of Conditions of Contract (with which we shall be dealing in Chapter 6) is given at Appendix 1.

Before studying any contract it is necessary always to refer to the list of definitions which apply to it. Among them will usually be found some which are not given here: this is because we do not have occasion to refer to those particular items in what follows. When reading a contract document, nouns with a capital (upper case) initial letter will always have the limited meaning given to them in the accompanying definition: with a lower case initial letter, they have their normal dictionary meaning. The use of a capital letter in the middle of a sentence thus has more than its usual significance. In some contracts (notably those relating to building construction) "the Architect" is used with the same meaning as we use "the Engineer" herein: what we say about the latter can equally refer to the Architect in that role.

The Employer: The body ordering the goods, works, or services covered by the contract.

The Contractor: The body whose offer to provide the goods, works, or services has been accepted by the Employer.

The Engineer: The engineer/architect appointed by the Employer to act on his behalf as provided in the contract. When a firm of consultants is appointed by the Client to manage a contract on his behalf, the firm itself will frequently be designated formally as "The Engineer" and will operate through a named individual whom they

	must appoint with full—or other specified—powers of the Engineer to act for and on behalf of the firm. The appointment is notified in writing to the Contractor and to the Employer.
Plant:	All machinery, apparatus, articles, materials and things of all kinds to be provided by the Contractor, other than Contractor's Equipment.
Contractors' Equipment:	Tools, tackle, stores, machinery, vehicles and other things brought to Site by the Contractor for the purposes of the Works but not for incorporation therein.
The Works:	All Plant to be provided and works and services to be done by the Contractor under the Contract.
The Site:	The place where Plant is to be delivered or work to be done by the Contractor and so much of the surrounding area as the Contractor shall be allowed for use in connection with the Works other than purely for the purposes of access to the place.
A Tender:	An offer by a tenderer to enter into a contract in response to and in general accordance with an enquiry from the Employer (or the Engineer on his behalf) at a price and on the terms stated.
Contract:	The bargain agreed between the parties in respect of the Works as expressed in the Contract Documents.
Contract Documents:	Documents of all types (including drawings) specified by the parties to a contract as recording their agreed intentions under the contract.
Contract Price:	The sum to be paid to the Contractor for the whole contract.
Contract Value:	The valuation on the basis of the Contract Price of a stated portion of the Plant or the Works in the condition and at the place in which it is at the relevant time.
Agreement:	As defined in Section 1.3 hereof.

The Employer and the Contractor are the two parties to the Contract. The Engineer is not a party to the Contract although he will normally have a well defined role in connection with it, specified in the contract documents. The Engineer has, in fact, what amounts to a dual role. Formally he is the appointed agent of the Employer authorised to act for the Employer in connection with the Contract on such matters and to the extent that the agreement between the Employer and the Engineer provides. His authority under the Contract stems from this appointment. However it is well established in law and recognised by the professional bodies in their Standard Forms of Contract Conditions that the Engineer must frequently act as a quasi-arbitrator between the two contracting parties and must be fair and be seen to be fair in his decisions, guided by the ethics and general practices of his profession.

Wherever in this volume we refer to the "Employer" taking a certain action the word is used in contrast to the Contractor or the tenderer. Employers will undoubtedly delegate the responsibility for many of the actions to the Engineer, depending on the exact terms of the agreement between them. Unless the circumstances clearly indicate otherwise, the likelihood of such delegation is to be read as implicit in "the Employer" even though the Engineer is not specifically mentioned each time. In establishing a contract between an Employer and a Contractor it will frequently be the function of the Engineer to draw up and issue the enquiry, to receive and evaluate tenders, to assist in the negotiation of points of difference between the Employer and the tenderers and to recommend to the Employer where the Contract should be placed.

In this volume we assume throughout that the Employer has retained a firm of consultants for the Contract and has duly nominated them in the contract documents as "the Engineer" with wide delegated powers.

Chapter 1
What makes a Legal Contract?

1.1 An agreement of views between the parties

A contract is formed when an offer (e.g. a tender) by one party is unconditionally accepted by the person to whom it is made—the second party. The keyword is "unconditionally" implying that full and complete agreement has been established between the parties. An acceptance accompanied by "if's" or "but's" or "provided that" does not form a contract but constitutes a counter-offer which in its turn requires unconditional acceptance by the first party.

The acceptance must be communicated to the offeror—silence does not mean consent. If it is sent by post it is considered "communicated" (and acceptance complete) at the moment it is posted and is not prejudiced by loss or delay in the post. Proof of posting is therefore important. An acceptance to be effective must be made within any limit of time stated in the offer or, if none is stated, within a "reasonable" time.

1.2 Essential Elements

In addition to the agreement evidenced by offer and acceptance there is a number of other essential elements of a contract. Some affect every contract: others are not likely to be met with in the normal run of engineering contracts such as we are concerned with in this volume. However for the sake of completeness all are mentioned. Some of the requirements cause a contract to be void automatically if they are not fulfilled. Others result in a contract which can be voidable (or partly voidable) as of right by one of the parties, usually the one who has been aggrieved or put at a disadvantage by the non-fulfilment.

The main requirements are:

- An *Intention* by the parties to create a legal relationship between them
- A *genuine consent* of the parties
- *Legal capacity* of the parties to enter a contract
- *Legality* of the objectives of the contract
- *Valuable consideration* passing between the parties

Each of these is now commented on in turn:

1.3 Intention to create a legal relationship

This requires that the parties fully intend that their agreement shall be one enforceable at law. Although a contract necessitates agreement, an agreement is not necessarily a contract. Two businessmen agreeing to meet at the Savoy for dinner have no intention of making the arrangement legally enforceable with penalties for failure—no contract exists. To distinguish between the two situations we have used in this volume the word "Agreement" (with capital initial) to indicate a formal contract drawn up in the terms of an agreement and concluded under hand or under seal (see Section 2.13 below). It should be mentioned that in many overseas countries it is customary to refer both to the bargain between the parties and to the formal Agreement by the word "Contract" with resulting confusion.

1.4 A Genuine Consent of the Parties

This phrase implies that the agreement must be clear, voluntary and without malice. It covers a number of matters as follows:

(a) *Fraudulent misrepresentation*

Fraud occurs when a party wilfully or recklessly makes a statement knowing it to be false with the intention of misleading the other party. Fraud can also occur by a party wilfully withholding material information or by some action conveying a false idea in relation to the contract, or by not taking reasonable care to ensure that a statement made is indeed true.

Fraudulent misrepresentation entitles the aggrieved party either to void the contract and sue for damages or to take advantage of the contract and also sue for fraud, claiming such damages as he has sustained.

(b) *Innocent misrepresentation*
i.e. the misrepresentor has been neither fraudulent nor negligent. A simple example, though not from the engineering field, would be if an art gallery sold an inexpert customer what they honestly believed to be a Picasso only to discover later that it was not. The aggrieved party cannot claim damages but only rescind the contract: however the courts at their discretion can award him damages instead of rescission.

Innocent misrepresentation puts an onus on the misrepresentor to show he did have reasonable grounds up to the time of the contract to believe that the facts he represented were true.

(c) *Duress*
Contracts can be voided by the weaker party if he can show he was forced to enter the contract by actual or threatened violence to his person—threats to his goods do not constitute duress. Happily this rarely affects engineering contracts.

(d) *Undue influence*
This extends the limited scope of duress to include the improper use of threats or influence acting for the benefit of the person making them. A contract so concluded is voidable by the weaker party on the grounds that *he did not accept its terms voluntarily*: he has the onus of proving that undue influence was applied by or in favour of the other party. The courts decide whether the influence he claims did affect the forming of a contract, and was "undue".

(e) *Genuine mistakes*
The effect of mistakes in drawing up a contract is involved and can only be briefly summarised here. There is no protection against mistakes at law or against errors of judgement and the parties are required to stand by the bargain they have struck. There are however certain mistakes of fact which can make a contract void or voidable, but as malice is by definition excluded, there is usually no question of damages being payable.

Important mistake as to the identity of the other party
For example an engineering company, a party to a contract, might unbeknown to the other party to the contract, have been taken over by another company unacceptable to the second party e.g. a competitor.

Mutual mistake as to the existence of the subject matter of the contract
For example the sale of an object which had meanwhile been sold elsewhere by an agent, or had been destroyed by fire, without either party being aware of it.

Mutual mistake as to the identity of the subject matter of the contract
For example an Employer negotiated with a contractor for the erection of a building having some alternative points of design. After detailed discussion the Employer decided to adopt the more costly alternative but this was not made quite clear in the final specification and the Contractor was able to show he had good reason for believing he had contracted for the other lower-priced alternative.

Under this same heading it should also be mentioned that a contract or parts of a contract may be voided by the courts on the grounds of uncertainty when the offending clauses are so vague or in such general terms that it is not possible to say what the parties intended by them when they signed the Contract.

Mistake as to the nature of the contract
When a person signs a document in the belief it is a fundamentally different one e.g. a contract to purchase believing it to be only an application to examine the goods on approval or return.

1.5 Legal Capacity of the parties to act

This usually covers matters which are not likely to affect engineering contracts of the sort we are concerned with, namely the fact that infants, lunatics, drunken persons and convicts under sentence are precluded or limited from concluding contracts.

There is, however, a similar matter which has a more practical significance for us i.e. whether a person signing a contract on behalf of an organisation has in fact been authorised to do so or is entitled by its regulations or articles of association to commit the organisation.

In the U.K. the requirement is that the person signing must be one whom the other party *has good reason to believe* would be so authorised: there is no requirement to check credentials except in cases of doubt.

Especially in contracts with overseas countries it is common practice to state in the document itself (either in the introductory recital or with the signatures at the end) the names and positions held by the persons signing for the two parties and a statement that they are respectively authorised to do so.

1.6 Legality of objects of the agreement

Certain contracts may be void by law because of the nature of subject matter e.g. wagering contracts, restrictive trading agreements, resale price maintenance agreements. Others, though not fundamentally void may contain a stated or implied requirement which is not legal e.g. to overload a ship, avoid payment of VAT, circumvent a regulation having the effect of law etc. The subject is a complex one, but if the illegal portion is separable from the legal, then in some circumstances the illegal portion only will be void.

1.7 Valuable Consideration

Although we are dealing with this essential requirement last, it is more likely to confront us in the engineering contracts we are concerned with than many of the foregoing points. We shall be constantly referring to it during the course of this volume.

It states that it is an essential element of every simple contract (that is, a contract not under seal) that a consideration of value must pass both ways between the parties. In effect this constitutes a bargain (as opposed to a gift) in which party A gives something acceptable to party B as a quid pro quo for the fulfilment of a promise by B.

Usually this bargain (in engineering contracts) is in the form of money in exchange for goods or services, but the consideration can equally be some right, benefit or interest accruing to one party or some forbearance, loss, detriment or responsibility suffered or undertaken by the other. We shall meet one or two such examples later in this volume. The consideration must be specific—vague or general considerations are not sufficient. It can be acquired at once or in the future, but past considerations are usually not sufficient. Provided a clear consideration is specified in the contract, the Courts are not concerned whether it is or is not adequate to match the obligations it relates to: the parties are held to the bargain they have made.

Certain special types of deeds (e.g. agreements and bonds under seal or guarantees as described in Section 2.11 below) are not subject to this "consideration" rule and are binding without it.

1.8 Contracts—verbal and written

1.8.1 The foregoing may have served to support a widely held—but mistaken—belief that a contract must always be a large and complicated document. This is not so: indeed almost every daily transaction involves a contract of one form or another. Purchases in shops, travel on public transport, eating in restaurants all involve contracts between the provider and the user.

For most matters the offer and acceptance forming the contract do not have to be in writing to be legally binding. The difficulty with verbal contracts is, of course, in establishing later exactly what was agreed or what the parties intended, and so for practical purposes in the engineering field we must consider contracts as being always in writing. The term "in writing" does not imply necessarily a deed or formal contract document: an exchange of letters can equally constitute a contract if they have the essential ingredients described in this chapter. What we have said about verbal contracts does however mean that engineers and others selling or negotiating a project, tender or order have to be careful not to promise, undertake or postulate anything verbally if there is no firm intention of carrying it out—they may find themselves committed to do so.

1.8.2 To complete the picture comment should be made on those contracts which cannot legally be verbal, but as they are mostly side-tracks from the main interest of this volume, the matter will not be pursued in detail. Briefly then there are two groups:—

(a) Contracts which must be completely in writing: these include share transfers, bills of exchange, promissory notes, marine insurance and certain hire-purchase contracts.

(b) Contracts which although they need not be completely set out in writing, must be "evidenced in writing" i.e. there must be one or more signed notes or memoranda recording all the material terms. They include three classes: contracts of guarantee, contracts regarding money lending and contracts relating to land. The last named includes not only purchase or lease of land but also contracts creating rights of way or of access, and such matters as fishing or boating rights. As regards guarantees see Section 2.11 of this volume.

1.9 Breach of Contract

Finally a few words on the subject of breach of contract (i.e. breach of the *terms* of a contract). The terms of a contract are legally considered to fall into two categories, conditions and warranties, which are differentiated as follows:

—A condition is a term expressing a matter basic to the Contract. A failure to perform its requirements implies a fundamental breach of an essential obligation undertaken.
—A warranty is a term dealing with a matter not vital to the essence of the contract, being subsidiary to the main purposes for which the parties entered into the contract.

Breach of a condition means the victim would be forced by the failure of the other party to accept something basically different from what he contracted for. The law then gives him the right, if he wishes, to treat the contract as at an end and to claim from the offending party all the damages he has suffered as a result. If he wishes he may instead choose to affirm the contract, but may still claim as damages the value of the reduced benefits he can expect to receive as a result of the breach. On the other hand a breach of warranty is regarded as being sufficiently unimportant in terms of the contract as a whole that it can be remedied solely by a payment in recompense for the disadvantages which the breach imposes on the victim: he is given no right to termination.

A typical example of a warranty is the clause found in many contracts dealing with the absence of defects in goods supplied, and undertaking to put right free of charge any that might appear in a stated period: indeed it is often referred to as "the warranty clause".

Clauses are however not always quite so straightforward. An undertaking by a party to complete the contract works by a stated date would, by itself, be a clear condition and not a warranty. Where the importance of keeping to the date is great, this is often underlined by specifying that "time" or "the completion date" is of the "essence of the contract".

However in some contracts, delay in meeting a date is associated with specified liquidated damages, payment of which by the Contractor is to be "in full satisfaction for any delay". By removing the right to terminate the contract this has the effect of reducing the clause to a warranty and no longer a condition.

Nevertheless if the delay should become much in excess of that provided for in the liquidated damages arrangement, the aggrieved party would still have a good cause of action for a fundamental breach of contract with the right to rescind the contract and to claim full damages suffered as a result (as with a breach of Condition). The period for which liquidated damages are prescribed is held to be the order of delay which was envisaged and accepted at the time the contract was signed. Anything much in excess of it was not countenanced by the contract or accepted by the parties, and thus represents a breach of an essential undertaking.

Chapter 2

What does an Engineering Contract usually consist of?

2.1 Object

A contract in the engineering field should ideally set out in writing exactly what are the agreed intentions of the two parties under any circumstances, however remote, which might arise during the performance of the contract. Clearly in practice a limit has to be set on the degree of unlikelihood of the circumstances which can justifiably be dealt with in any particular case. The basic objective must be to avoid a cause for argument when things go wrong and relations between the parties have become strained. It is better to agree what is to happen if such circumstances should arise by discussions in a calmer atmosphere beforehand. Minor contracts will generally justify leaving more unsaid since, by virtue of their relative simplicity, they offer less risk of trouble arising. However in every case those points that are included must be covered precisely and unambiguously so that the executives of the parties and indeed, in extremis, the courts themselves can interpret them in the exact way the parties originally intended.

The good contract draughtsman does not seek to "pull one over the other side" by including provisions the other party may fail to appreciate until it is too late. Rather he seeks to settle difficulties at the negotiating stage so that the document will be conducive to harmony between the parties during the performance of the Works, avoiding the misunderstandings and undefined areas of responsibility which are so fruitful of disputes later. It must not be regarded as an instrument for extracting as much as possible out of the other party.

2.2 Contract Documents

A contract should always specify what documents, letters, specifications, etc., are to be regarded as the "contract documents", and these normally define the contract. If however in a court of law the contract documents themselves leave uncertainty as to the true intentions of the parties at the time of signing the contract, other letters and documents might, subject to the rules of evidence, be introduced by either party as evidence of such intentions. In the sort of contract we are considering the contract documents will usually comprise some or all of the following:

- A statement of the scope of the Contract
- Data (e.g. site conditions, climate) affecting the execution of the Works
- A Technical Specification
- Bills of Quantities (especially in civil engineering and building construction contracts); Priced Schedules of Plant (in plant supply contracts)
- Programme for the construction/completion of the Works
- The Conditions of Contract
- Site Regulations
- The Form of Tender: the offer
- Unconditional acceptance of the tender
- Guarantees or Bonds
- A formal Agreement
- Additions or variations of the above made during negotiations or subsequently by agreement between the parties

We will look at each of these separately:

2.3 Statement of the Scope of the Contract

The statement of the scope of the Contract is made by the Employer in inviting tenders. It introduces the subject of the Contract and explains in broad terms what the Employer will require by way of Work or Plant from the successful tenderer, and by when he will want it to be complete or in operation.

2.4 Data affecting the execution of the Works

Data affecting the execution of the Works is issued by the Employer with the invitation to tender and includes such things as Site geological data, access, limitations on working hours, local conditions, other work being carried out on the Site, machinery, services and supplies which the Employer will provide for the Contractor's use, and so on. All such points may affect the tenderer's costs and hence his offer.

2.5 The Technical Specification

The technical specification of the Works is, as the name implies, the detailed requirements of the Plant or the Works and their operation.

It will usually include some or all of the following:

- Descriptions of the Works
- Design information for the Works
- Drawings including layout and Site plans
- Plant operation, maintenance and training requirements
- Inspections and tests on acceptance of the Works
- Performance standards to be achieved
- Test procedures

It is usual in civil engineering and building contracts for the Employer to specify the job in full and complete detail. For plant supply contracts the specification is more usually a functional one leaving the tenderer free to use his specialist design know-how or available range of plant to the best advantage. In either case the more precise the Technical Specification the more smooth running will be the execution of the contract Works and the more obvious the point at which the Contract requirements have been adequately fulfilled: in the absence of a precise specification the Employer always tends to expect more and the Contractor to give less.

2.6 Bills of Quantities, Prime Cost Items and Provisional Sums

2.6.1 Bills of quantities are usually issued by the Employer (or the Engineer on his behalf) with the invitation to tender. They itemise the various supplies and activities comprising the enquiry and the different services for which the Employer requires broken down prices from each tenderer. For each item of the bills the Employer indicates the quantity which he considers will be required by the Works and the tenderer inserts for each item his appropriate price and the unit rate on which it is based. In civil engineering and building contracts (where the Employer normally draws up the detailed design of the Works) the bills of quantities will, of course, cover all the elements which make up the Works so that totalling the prices for the individual items can lead directly to a lump sum tender price. In some cases (especially in connection with building construction) the description of the items in the bills can often be so specific that, taken in conjunction with the drawings of the Works, they form a full specification of the Employers requirements and no separate specification document is needed.

The Employer's estimates of quantities are obtained from the specifications and drawings relating to the Contract. If the design is insufficiently complete by the time it is necessary to go out to tender to enable this to be done with reasonable accuracy, competitive pricing of the various categories of work involved can nevertheless be called for from tenderers either by using approximate quantities of the right order, to give the tenderers an idea of the size of the the job or, if even this is not realisable, replacing the bills of quantities by a schedule of rates. In the latter case rates of charge only are called for and no attempt made to evaluate even an approximate lump sum tender price from them.

If the quantities in the bills are known to be complete and

probably firm, unforeseen changes during the course of the contract can be dealt with as contract variations and the Contract Price adjusted accordingly. In all other cases, the final Contract Price is determined by remeasurement of the Works (see paragraph 7.1.5): the tender documents must in every case make it quite clear which procedure it is intended shall apply.

Besides allowing the Employer to make more detailed comparisons of the tenders received, bills of quantities establish in a remeasurement contract such as we have described a "unit rate" for every item which can be used to evaluate the Contract Value of the actual quantity which has in the event to be provided in carrying out the Contract. The unit rates are also useful in a fixed price contract not subject to remeasurement for assessing (whenever they relate to similar items) fair prices for contract variations which have to be introduced during the course of the Contract.

2.6.2 A Bill of Quantities will also include any *Prime Cost Items* the Employer may specify. These refer to items of plant or services which the Employer intends shall be sub-contracted to suppliers of his own choosing. They might refer, for example, to specialised proprietary equipment or to items for which the Employer has already issued an enquiry and received tenders. Prime cost items specify to the main tenderer the equipment or services to be provided and the Employer's assessment of how much the Contractor will have to pay the manufacturer or supplier ('the nominated sub-contractor') for them. The tenderer is expected to add a percentage to this prime cost as his handling charge and for his profit, and the amount thereof is either specified by the Employer in his invitation to tender, or more usually left to the tenderer to quote in his tender. The prime cost sum ultimately included in the Contract is the sum certified by the Engineer as actually payable to the sub-contractor. If this should differ from the anticipated prime cost, the Contractors cover is adjusted pro-rata. Prime cost items are always regarded as a part of the Contract and the Contractor can rely on their expenditure. The value to the Employer of prime cost working lies usually in one or more of the following, but it often brings contractual complications in its train (see Section 8.3 below):

- It ensures that a particular item (for example a particular brand of proprietary article) will be used in the Contract.
- It ensures that specialist work which is considered to be beyond the competence of the Contractor will be designed and/or constructed by specialists under the control of the Engineer and Contractor.
- It gives the Contractor an indication of the size and complexity of the specialist work associated with the Contract enabling him to allow for it in his programme and in his provisions for supervision and co-ordination.
- It enables the Employer to put in hand procurement of long-delivery specialist items in advance of the full definition of the Works and of the whole procedure leading to the placing of the main contract.
- It enables an Employer to take advantage of discounts he may be allowed which would not be available to the Contractor.
- It is sometimes used when a specialist item which is to be used still lacks detailed definition possibly due to discussions with or development work by the intended supplier still being in progress.

2.6.3 Bills of quantities will also include any *Provisional Sums* which the Employer wishes to be included in the tenderer's price for items which cannot be envisaged accurately at the time of tender, or for possible later extensions to the scope of the Works or the technical specifications which he may wish to introduce, or to meet extra quantities he realises may become necessary as the Contract proceeds. By allowing a provision for the cost of all such items at the tendering stage the Employer ensures they are included in his budget for the Works.

Unlike prime cost items, the option of whether a provisional sum is to be expended or not rests entirely with the Employer, and the Contractor has no grounds for complaint or claim if it is not released and never included in the Contract Price. By including provisional items in the tender the Employer also ensures that incompletely defined or possible extensions of the Works are associated with the Contract from the start and any subsequent disagreement with the Contractor as to whether they are of such a nature that they can legally be brought in by a written variation to contract is avoided. Further information on bills of quantities contracts is given at paragraph 7.1.5 below.

2.7 Programme for the construction/completion of the Works

2.7.1 This section of the contract documents records the critical dates during the progress of the contract Works expressed either by calendar dates or by elapsed weeks after a start date for the Works, to be determined subsequent to the tender date. The dates may be inserted by the Employer as the dates by which he requires the Works (or sections of the Works) to be completed and any critical points during the execution when the progress has to tie in with other related contracts. Alternatively the tenderer may be required to insert his best promise of achievement as part of the competitive aspect of his tender. The latter is more commonly met with in conventional building contracts and in plant supply contracts, when best delivery and completion dates are called for.

Whichever way it is done in any particular tender, the document establishes the contractual dates to which the Contractor will be bound, and if he fails to keep them the Employer may be entitled to claim appropriate damages. These may be the full amount the Employer can show, at the time, he is going to suffer as a result of the delay, or they may be predetermined and agreed by the parties at the time of tender and written into the Contract as liquidated damages.

Liquidated damages so agreed may not exceed in value a fair and justifiable estimate by the Employer (at the time of tender) of what he could reasonably expect to suffer in the circumstances of the delay. Any greater sum than this would represent a 'punishment' of the Contractor for being late, and as such is not legally permissible. A sum less than the estimate is permissible as long as the Employer is prepared to accept it and the Contractor agrees.

If, however, liquidated damages 'in full satisfaction for the delay' are not specified in the Contract, or if the length of the Contractors delay exceeds the period (if any) for which liquidated damages are prescribed to run, or is unreasonably long, then (as we said in Section 1.9 above) the Employer may become entitled to terminate the Contract on the grounds of breach of an essential condition, in addition to recovering his appropriate damages.

2.7.2 In the more involved projects (especially in the constructional field) it is usual to require the tenderer to submit with his tender an outline programme showing his estimated timings for the major stages in the progress of the Works, in order to show that he has a workable plan to meet the contract dates. After the contract is placed the Contractor is generally required to draw up within a few weeks fully detailed plans for the execution of the Works (in the form of network plans or suitable charts). These are submitted to the Engineer and, after approval, are used to establish and regulate the day-to-day progress being made by the Contractor.

2.8 The Conditions of Contract

2.8.1 The conditions of contract are the 'Rules' by which the contract is run. They set out the rights and obligations of the parties and agree the action that will be taken by the parties if various eventualities arise during the course of the Contract. The range of eventualities they cover varies from case to case depending on the nature of the contract and the views of the parties but even for small contracts the conditions of contract may run to several pages of print. For really large contracts they may extend to several volumes.

Contract conditions form much of the legal basis of the Contract on which any ruling by the courts would be made. Consequently they need to be written with care and precision so as to be clear and unambiguous. Luckily many of the points to be covered are common to a large number of types of contract and, in addition, experience has led to a form of words which is generally acceptable to both parties. As a consequence there exists a number of standard forms of conditions of contract which have fairly wide recognition in the U.K. and some which are recognised internationally. The use of these can remove a lot of the need for composing new conditions for each and every contract. We say more about standard forms of conditions of contract in Chapter 6 below.

2.8.2 The conditions which it is intended shall apply to any particular contract may be proposed by either party. With formal tendering for large contracts, suitable conditions are usually nominated in the tender enquiry and the tenderer is expected to accept them. In many cases they will be basically one of the standard forms of conditions of contract with a minimum number of additions or modifications which the Employer or the Engineer considers necessary in order to make them suit the particular project or enquiry. The additions are frequently incorporated under a separate heading "Special Conditions of Contract" and the modifications by a list of amendments to the standard form concerned.

The tenderer in making his offer may propose further changes on which he makes his offer contingent or, more extremely, may propose the substitution of a quite different set of conditions, either his own or another standard form with which he has had satisfactory experience in the past and considers more appropriate to the contract for which he is asked to tender. Such an occurrence is more likely to be encountered with plant supply contracts where the Contractor usually has the responsibility for design and the work is conditioned by his manufacturing process. These raise many contractual points which are not met in contracts for the construction on the

Employer's Site of works for which the Employer has full design responsibility. If the tender is otherwise an attractive one, negotiations must then take place until conditions are achieved which the Contractor and the Employer (with the advice of the Engineer) both consider they are able to accept.

2.8.3 The selection of the most appropriate of the standard forms of conditions of contract available for use with any particular contract calls for a detailed knowledge of their contents. The incorporation of any negotiated changes in a standard form is a skilled job: it is not only a matter of arriving at the right arrangement of words to express unequivocally and clearly in the appropriate clause the new situation, but also of appreciating what consequential modifications elsewhere in the standard form the changes will bring in their train. Contracts involving two or more engineering disciplines (e.g. mechanical and civil) pose their own special problems as they introduce differing basic principles of operation.

2.8.4 Legally the absence of agreed conditions of contract does not invalidate an otherwise valid contract. It means that the chances of disagreements between the parties later are increased and the chances of resolving such disagreements by mutual negotiation (at a time when relations are probably strained by the contract having gone wrong) are reduced. If the contract relates to the sale of goods for money, the rights and obligations of the seller and purchaser are established in general terms by the Sale of Goods Act (1893) and its recent amendments in the Supply of Goods (Implied terms) Act (1973). Matters such as delivery, transfer of ownership and risk, acceptance and rejection of the goods, quality of the goods, payment, warranties and the like are well covered. The Sale of Goods Act does not however cover contracts for work, labour or services (as the title implies); neither can it deal with the special or detailed aspects of any particular contract. Settlement by recourse to arbitration or the courts is all that is then left and both procedures are apt to be slow, involved and expensive.

2.8.5 When drafting or modifying contracts, great care must always be taken not to use terminology which is open to misinterpretation even if it is known that all concerned at the time fully understand what is intended. Contracts frequently have to be interpreted by others at a much later date (and sometimes, in an extreme case, by the Courts) and ambiguities only lead to trouble and dissatisfaction. As an example, in one contract a considerable claim and interminable argument arose from a 'misinterpretation' by the Contractor of what was embraced by "System" in the term "the Electrical System...".

All terms which are not of themselves unambiguous must be clearly defined. Having once been defined, the word must thereafter be spelt with a capital letter and used only with its strict defined meaning.

2.9 Site Regulations

As distinct from the conditions of the contract, which are matters vital to the operation of the Contract by the two parties, Site regulations deal with rules which, though of lesser importance, the Employer wishes to impose on the Contractor for the efficient running of the Contract at the Site. They deal mostly with operational and administrative matters such as:

- Traffic routes to be followed
- Site cleanliness
- Fire precautions
- Labour relations
- Noise suppression
- Welfare arrangements: sanitation and the like.

The wording of the tender offer or the form of tender is such that the tenderer undertakes to observe these regulations and to require his sub-contractors to do likewise.

2.10 The Tender

2.10.1 The tender forms the 'offer' put up by the tenderer for the Employer's eventual 'unconditional acceptance'. It is important in most cases to put a period of validity on a tender after which it is no longer open to acceptance. After the expiration of the period of validity, if the Employer has not accepted the tender, the tenderer is at liberty either to amend it in any way he wishes (including revision of price or delivery date) with a new validity period, or to revalidate it as it stands for a further stated period, or to withdraw it altogether.

In many countries (including the U.K.) a tender can also be withdrawn at any time during its validity if it has not already been accepted by the Employer and the phrase "open to acceptance...unless previously withdrawn" is often incorporated to emphasise the right. Problems can occur if, for example, an acceptance and withdrawal cross in the post.

An Employer is usually upset by tender withdrawals whilst he is busy appraising them and he is under no obligation whatever to consider any substitute tender which may be proffered. Indeed he should refuse to do so as it could well be based on a leakage of information about the size of competitors offers. Some Employers include in their form of tender an undertaking to be signed by the tenderer that he will not withdraw his offer during the validity period: others state that withdrawal will influence their inclusion of the tenderer in future enquiries. Where it is sufficiently important to ensure non-withdrawal of an offer during its validity period, the tenderer can be invited either to provide a formal undertaking under seal (a tender bond) or to enter an agreement not to do so. In the latter case the Employers "consideration of value" might be the fact of his having given the tenderer the opportunity to tender, provided this is acceptable as such by the tenderer.

2.10.2 The tender will state the Conditions of Contract to which the offer is subject. If these have been nominated by the Employer in his enquiry they may not be fully acceptable to the tenderer who must list in his tender the changes that he deems necessary. The tender will also state, in a form appropriate to the type of work concerned, the Contract Price and the terms, currency and method by which the various payments are to be paid by the Employer.

2.10.3 If, as is usual in engineering work, the technical specification for the contract has been issued by the Employer with his invitation to tender, the tenderer must include in his tender:

- a schedule of all points in which the Plant or the Works he is offering do not conform to the specification issued.
- additional technical information concerning his offer to augment the technical specification as issued.

Where the latter is a purely functional one (as may be the case in tenders for the supply of plant) the "additional information" may well comprise the full technical description with general arrangement drawings and performance charts of the plant put forward by the tenderer to fulfil the specified functions.

2.10.4 Provided that he makes at the same time an offer fully in accordance with the technical specification issued, it is usually considered acceptable for the tenderer to submit an alternative scheme of his own if it seems to him to be a superior (or cheaper) method of achieving the Employer's desired result, for example when the Employer's design concept is not well suited to the tenderers existing equipment or traditional methods.

2.11 Bonds and Guarantees

2.11.1 Without entering too deeply into the precise legal situation (which can be complex) we may define for our present purposes:—

(a) *A Deed* is a general term for a contract document under seal (see also Section 2.13 below).

(b) *A Bond* is a deed in which one party promises to perform something under stated circumstances in a specified way for the benefit of a second party.
If the undertaking is to pay money it is called a "common money bond"; if it is to produce some act or event it is called a "bond upon condition".
A bond only needs written acceptance by the party benefiting to become binding. It is discharged and becomes

void as soon as the undertaking has been performed as specified.

(c) *A Guarantee* is a promise by a third party "C" to be answerable to a party "A", in a manner and to an extent specified, for the payment of a debt or for a default or miscarriage in the performance of a duty owed by a party "B" to the party "A", in the event that party "B" fails to perform his engagement.

Note that the third party (the *"Surety"* or "the *Guarantor"*) pledges himself to act only if party "B" has failed to perform his duty.

A guarantee does not have to be a bond (i.e. a formal document under seal) but in engineering contracts it is often so required in order to emphasise the validity of the actual words written (see para 2.12.1). A guarantee must however (at any rate in most of the cases met with in engineering projects) be either in writing or "evidenced in writing" i.e. recorded as to its major features in a memorandum of some sort.

In the commercial and domestic fields of buying and selling, a guarantee is rarely, if ever, in the form of a bond. Indeed many so-called "guarantees" given by manufacturers for their products are in fact warranties since no third party is involved as surety.

2.11.2 There are four main types of bond or guarantee in common use in connection with engineering contracts, some or all of which may be called for by an Employer in his enquiry:—

(a) *A Tender Bond* (also called a "Bid Bond")
This is a bond called for by an Employer to accompany a tender in which the tenderer undertakes to maintain his offer unaltered and open to acceptance during the full validity period stated in it and to accept any contract based on the tender which may be awarded to him during that period. If called for as a guarantee, then the tenderer must produce a similar promise from his bank or insurance company or other surety (or sureties) acceptable to the

Employer. If the tenderer defaults, he or his surety (as the case may be) will pay the Employer as damages the full amount stated in the bond. Tender bonds are not usually called for in the U.K., but are quite common in tenders for contracts abroad.

An alternative to a tender bond sometimes used is a tender deposit i.e. a sum of money deposited with the Employer or placed at his disposal at the time of tendering which the tenderer forfeits if he defaults in a like undertaking to the one above. A tender bond may typically be for 1% or 2% of the tender price.

(b) *A Performance Bond* (also known as "Security for Due Performance", "Contract Guarantee" etc.)

This is an undertaking given by the Contractor when the Contract has been awarded to him that he will "punctually truly and faithfully perform and observe all his obligations under the Contract". For obvious reasons, this bond is usually required by the Employer in the form of a guarantee by an approved surety. Performance usually covers not only technical completion but all other obligations (tests, training, programme of contract dates, payment of any monies due, repair of defects etc.). It also covers non-performance as the result of bankruptcy, commercial take-over and the like.

In the event of a default or breach the Employer can demand from the surety payment of such sum as represents the damages he has sustained up to the limit of the sum guaranteed. The claim may be the full amount of the guarantee but if not, the balance remains in being until the contract is duly completed, and may be subject to further claims in the event of any further breaches.

Performance bonds are common both with home and overseas contracts and may typically be for 10% of the Contract Price. The surety (especially if a bank or insurance firm) frequently refuses to involve itself in the validity of the claim of breach, and will pay up against the certificate of the Engineer or even "on the first demand of the

Employer"—especially in some foreign contracts—without further question or investigation.

In contracts for the supply and erection of plant, a performance bond may also cover the efficiency of operation of the plant as installed (see *Plant Performance Bond* para (d) below), but the latter is often kept separate with the Contractor himself being liable on a sliding scale for any deficiencies in operation.

(c) *A Repayment Bond* ("Advance Payment Bond"; "Progress Payment Bond" etc.)

In contracts in which the Employer agrees to pay part of the Contract Price in advance of receiving its worth in goods or services, he may require from the Contractor a bond or a guarantee undertaking repayment in the event the money remains unearned. This might apply to payments 'with order', progress payments prior to delivery, final payment up to 100 percent before completion of the maintenance period and such like. The Contractor (or the surety, as the case may be) forfeits from the bond any sum which remains unearned up to the maximum stated in the bond (which is not necessarily the same sum as the part-payment of the Contract Price concerned).

(d) *Plant Performance Bond*

In contracts in which the Plant after installation and tuning up is required to pass tests demonstrating, for example, a specified efficiency or a specified quality and/or rate of output or the like, the Employer may (and usually does) require a stated sum or sums of money by way of damages if the Plant fails to achieve such performance. Not unusually these sums are on a sliding scale depending on the degree of failure. Often the Employer in this case is content to bind the Contractor by a clause in the contract that he will, if he fails to meet the required performance, forfeit specified sums from payments of the Contract Price due to him, but if the extent of the damages the Employer stands to incur by such failure is much greater than the part of the Contract Price still unpaid, he must arrange to recover through a bond, or a guarantee by a surety to be approved by him.

Plant performance bonds are frequent sources of dispute between the parties and, particularly as regards the details of the tests themselves, require a great deal of care in drafting. The following points at least must be considered and covered either in the bond or (more usually) in the test specification in the Contract:

—Exactly what performance figures are promised and what tolerances may be accepted without penalty?

—Under what parameters of plant operation are the tests to be made? What are the acceptable limits on raw materials used during the tests and on services provided (e.g. composition, purity, water flow, or electrical voltages)?

—How are variations outside limits during the course of the tests to be treated?

—What parameters must be measured? How and by whom?

—Who provides the measuring equipment and meters and how is their accuracy verified?

—Who operates the Plant during the tests and what rights has the Contractor to direct its operation? Can contractor elect to discontinue a test?

—Who controls the test, supervises measurements and keeps records?

—Minimum duration of test? Hours per day? Continuous or intermittent run? Warming-up period? What forms of breakdown or interruption can be ignored?

—How and by whom are results assessed?

—In event of failure to reach guaranteed performance, what right has Contractor (i) to adjust Plant, (ii) to redesign and modify (iii) to replace Plant? How many attempts to pass tests may be made before the penalty is imposed?

—What are the consequences of failure? Under what conditions may Employer (i) reject Plant and terminate Contract, (ii) retain Plant with specified reduction in price or payment of damages specified, (iii) require due performance of Contract, i.e. Contractor to continue work until successful?

2.11.3 Bonds and guarantees of the types described are not parts of the contract documents of the main engineering contract. They form, in effect, separate "contracts" on their own. Thus a tender bond (para 2.11.2(a)) is binding from the date the tender is submitted and is discharged by the signing of the main contract or the expiry of the validity period of the tender. The other three examples (b) (c) and (d) can only be drawn up *after* a contract has been concluded. And in the case of a guarantee, the surety giving the undertaking is not a party to the main Contract. A bond is normally related to a specified contract i.e. in the form the latter exists at the time the bond is signed. Any subsequent variations to the contract can easily have the effect of making the bond void unless special precautions are taken in the wording of the bond to obviate this. The same applies equally to guarantees.

To place a contractual obligation on the Contractor to produce the post-contract bonds, there must be a condition in the contract stating exactly the form and extent of each bond required, and specifying a period after the contract date during which they must be produced. The wording of the bonds and also the identities of the sureties themselves must be specified as being "to the satisfaction of the Employer "(or the Engineer, or both), and usually the categories of individuals or parties which will be acceptable as sureties are specified. Banks and insurance institutions are frequently specified and these often have very definite ideas as to the terms of any bond they are prepared to enter. They may allow little or no deviation from their standards. They undertake to pay only on presentation of a document specified in the bond, which they always accept at its face value. Sometimes, especially in some overseas countries, the document need be no more than a statement of entitlement from the Employer himself (which affords little protection to the Contractor!), but more usually the authority for payment is a certificate of entitlement by the Engineer in his role of unbiased mediator. This certificate should state that a breach of contract or default by the Contractor has occurred entitling the Employer to certain sums as damages in accordance with the terms of the bond.

2.11.4 The restitution afforded to the Employer in the event of default by the Contractor is usually in terms of money but is not necessarily so. For example a not uncommon form of performance bond requires a parent company to undertake as surety for one of its subsidiaries the specific performance of the obligations under a contract in the event of default by the subsidiary.

2.12 A Form of Agreement

Although under English law a valid contract may be formed purely by documents constituting an offer and an unconditional acceptance, it is sometimes necessary or advantageous to embody or confirm the contract by drawing up a formal Agreement.

The standing orders of many public bodies, (for example local and county councils) require that all contracts over a certain sum are so embodied in a formal Agreement which has to be signed and sealed as a corporate document on their behalf by certain officials they have empowered to do so.

Confirmation of a contract in a formal Agreement can be of value:

> —To emphasise the formality and importance of the Contract.
>
> —To summarise the highlights of a complex contract.

- To stress that the Contract is with the Employer in cases in which the enquiry, tenders, negotiations, and correspondence have all been handled by the Engineer (in his role as agent for the Employer).
- To schedule the items forming the contract documentation.
- To tidy-up a contract situation when there have been extensive negotiations following the tender, and the contract documentation and details have suffered involved modification.

The Agreement, besides listing the contract documents, normally sets out the main features of the Contract, e.g. the agreed Contract Price and its terms of payment, completion dates, guarantees and bonds, agreed liquidated damages and suchlike.

To make the subsequent execution of an Agreement obligatory on the parties to a contract formed by offer and acceptance, such requirement must be included as one of the Conditions of Contract. This Condition will usually specify a period within which the Agreement must be drawn up by the party nominated so to do and completed by the other. If the former fails to comply, the latter may determine the Contract owing to breach of condition. Instead of requiring an obligatory Agreement, the condition is often written to give the Employer the right to call for an Agreement at his option within a stated period and to make the Contractor responsible for signing (or sealing) such an Agreement if the Employer exercises his right.

2.13 Agreements 'under hand' and 'under seal'

2.13.1 Agreements of the sort we are considering can be executed in one of two ways, either 'under hand' (i.e. by the signatures of the two parties) or 'under seal' in which the seals of the two parties are affixed to the document. In practice the difference between the two forms as far as concerns our present purposes is:

- In an Agreement under seal statements of fact cannot subsequently be called into question.

- The period of limitation for starting a legal action or arbitration proceedings is extended to 12 years (in the case of an Agreement under seal) from 6 years (for an Agreement under hand). This may be of great importance in contracts in which defective work may not become apparent for some years.

- Documents in which a consideration of value does not pass both ways between the parties (see Section 1.7) must be under seal.

The second of the foregoing points, though simple in statement, conceals considerable complexity in its interpretation, namely in establishing the instant in time from which the two periods start to run. Legally it is from the date "a cause of action accrues" either by a breach of contract or the performance of a civil wrong (or "tort"), and in practice establishing this can often be a matter of legal argument and judicial ruling. Depending on the exact circumstances of the case and the terms of the contract it can, for example, be the point at which the basic cause occurred, or the date on which the Employer took over the completed Works, or the date of the final certificate following the end of the Contractor's maintenance obligations. Indeed in a recent appeal, the House of Lords ruled unanimously that where damage occurs in a building as a result of faulty foundations the cause of action may, in certain circumstances, not "accrue" until the damage itself becomes manifest, even though this may be years after the Contractor has finished work and left the site. The period of limitation is, in effect, postponed indefinitely.

2.13.2 Agreements under hand require only the signatures of the parties, each duly witnessed by one witness (who also signs and states his occupation). For Agreements between corporate bodies, the signatory on behalf of each body must be an official properly authorised by the rules of that body to commit it in the way and to the extent specified in the Agreement. In the case of a limited company its Articles of Association usually cover this point (for example they frequently specify 'any of its directors') but see also Section 1.5 above.

2.13.3 Agreements under seal between corporate bodies are executed by each party fixing or embossing at the foot of the Agreement, its corporate seal. In each case a witness authorised by the Articles of Association of the body identifies the seal and signs to its having been affixed in his presence. When the parties are individuals or have no corporate seal, an Agreement is executed 'under seal' by the signature of duly authorised individuals as already described (and their witnesses) after the formula 'signed, sealed and delivered'. It is customary to affix an adhesive red disc as a token seal against the signature to draw attention to it being 'under seal' and not simply 'under hand'.

Agreements executed under seal require a revenue duty stamp to be affixed by law generally within 30 days after the date of execution thereof. Failure to do so does not invalidate the agreement but could lead to trouble in a court of law as the Stamp Duty Act has not been complied with. These stamps are embossed on the Agreement by taking or sending it to an Inland Revenue Stamp Office and paying the appropriate duty. There are eleven such offices distributed in the principal cities of England to which documents may be taken for immediate attention, including two in London (Bush House, Strand, WC2 and 61 Moorgate, EC2). Small parcels of documents can be handed in with the amount of the duty at a main post office, for onward transmission. On completion, the documents are returned through the normal postal service. Alternatively documents may be sent direct by post to a special stamp office in Worthing, Sussex. The amount of duty depends on the subject matter of the Agreement but for all Agreements dealing with the provision of goods or services of the sort with which we are here concerned, the duty at the date of publication of this volume is 50 pence.

Agreements under hand do not require a stamp.

2.14 Subsequent Variations to a Contract

Variations to an existing contract fall normally into two categories:

— Variations to the specification or extent of the *contract Works.*
— Variations to the *terms and conditions* of the Contract.

2.14.1 Most of the standard forms of conditions of contract (para 6.5) contain a so-called "Variation Clause" which empowers the Engineer to amend within wide limits the specification or the extent of the Works. Some, in addition, protect the Contractor's interests in this matter by limiting the total value of variations the Engineer is entitled to make to a specified amount (for example ± 15 percent of the Contract Price). Beyond that limit, variations of the Works can only be made with the agreement of the Contractor. Even where there is no expressed limit (for example in the Standard Conditions of the Institution of Civil Engineers) the Courts would be unlikely to support an Engineer who sought to introduce variations which either substantially changed the character of a contract (up or down) or had no direct bearing on the scope of the original contract.

2.14.2 Variations to the terms and conditions of a contract can be made at any time by agreement between the parties thereto in the form of a further contract expressing the changes. If the latter are extensive, it is usually better to rescind the earlier contract and replace it by a new one. Such contracts of variation or rescission have to follow the normal rules for contracts we have described elsewhere in this volume, and these include the need for a consideration of value to pass both ways between the parties. Contract changes are often such as by their nature benefit both parties, but if only one should benefit, he would have either to forego some advantage under the contract or to give some new service or pay a sum of money for the benefit of the other. Contracts of variation or rescission are normally made under hand or under seal in conformity with the original contract to which they relate.

2.14.3 It must be noted that the powers of the Engineer to vary the Contract by virtue of his appointment are strictly limited. He may amend the specification of the Works within the limits set by the variation clause, but he has ex officio no powers under that or any other clause to alter the terms and conditions of the Contract. The Contract is between the Employer and the Contractor: the Engineer is not a party to it.

Of course the Engineer like anybody else can be formally appointed by the Employer ad hoc as his agent or given power of attorney to negotiate and conclude on behalf of the Employer an amendment to Contract with the Contractor. It is a situation which needs careful watching to avoid the agent himself assuming liability and to ensure a legally sound contract results. We return to the subject in Chapter 4 below.

Chapter 3
Negotiations leading to a Contract

3.1 Need for Negotiation

It will be clear from the foregoing chapters that a contract cannot always be expected to emerge directly from an offer and a straight acceptance, except possibly in the case of relatively simple works contracts which have been fully specified in the tender documents. In large or multi-disciplined projects there may be several periods of discussion and negotiation between the Employer and the preferred tenderer or tenderers before an identity of views, which can be confirmed in a signed contract, can be reached.

3.2 Enquiries against firm Specifications

The first steps in this direction may well be met as soon as the invitation to tender is issued, when tenderers will have other views on how the technical needs should be achieved or on the contract conditions. However, most such contracts are the subject of competitive tender and it is basic to such tendering that all offers are made on a common footing. The Employer cannot, therefore, at this stage agree any modifications to specification or conditions with one tenderer *unless they are promptly notified to all the others,* with sufficient time for them to be able to take the revised basis into their consideration.

(It is for the same reason—to preserve the comparability of tenders—that tenderers wishing to quote against their own alternative specification are usually required to quote as well against the Employer's specification as issued.) It is wiser, if it is at all possible when dealing with a tender having a firm specification, to avoid any discussion of detail with tenderers once the tender documents have been issued, as it is difficult to be sure no unfair advantage is being given to the tenderer concerned.

3.3 Enquiries regarding large and complex projects.

3.3.1 Preliminary tenders

In the case of large or complex projects such as, for example, "turnkey" or "process" projects an enquiry may have to be issued initially against a preliminary functional specification partly on account of the time factor and also to harness the expertise of the tenderers themselves in drawing up a firm specification which will both yield the best results for the Employer and at the same time be consistent with the views and methods of the firms who may have to execute the work. One approach might be to use a *Cost-plus* or a *Negotiated Contract* (see paras. 7.1.3 and 7.2.2) with a firm whose reputation seems most appropriate, or (if design is sufficiently advanced) to select the firm by a preliminary 'competitive' enquiry calling for estimated prices. The Employer keeps open his options during the negotiations which follow, to change to a different contractor if the first choice proves unsatisfactory, but the time factor may make this difficult in practice.

The negotiated contract or cost-plus contract procedures have the advantage that the expertise and full co-operation of the chosen contractor can be expected during the development of the project from its preliminary to its final state. They have however, the disadvantage to the Employer of being non-competitive once the future contractor has been chosen. In the particular case when the specification is reasonably settled at the time of the preliminary tender, if only in outline, some

51

degree of competitiveness can be introduced by a *Target-Cost Contract* (see 7.1.4 below) but this will not be attractive to Contractors unless a closely representative target cost can be estimated at the enquiry stage.

3.3.2 Two-stage tendering

A more competitive basis can be achieved by a two-stage selection process. There are a number of variations on this system of which we will refer to three:

(a) The 'basic' method: For the first stage tenders are called for against a preliminary functional specification, or against alternative provisional specifications, tenderers being expected to give sufficient explanation of their detailed proposals to show that they are practicable. From the resulting offers a single 'optimum' specification is selected by the Enployer and re-issued for fully compliant tenders by all tenderers. The method has obvious disadvantages from the point of view of the tenderers, not the least being the cost of the abortive work in developing and pricing schemes which are subsequently dropped. The chosen design may well satisfy none of them.

(b) A variation of (a) in which the provisional enquiry is issued for a short period of study by tenderers but no bids are called for. Instead, a general meeting of all potential tenderers is held to criticise, discuss and amend the initial enquiry or to propose new alternatives. All aspects of the enquiry, technical, commercial and contractual, are covered and, if called upon by individual tenderers, separate meetings can be held with them to clarify points of uncertainty. The preferred features and acceptable compromises are then embodied into a single firm enquiry which is issued formally to all tenderers for fully-compliant bids.

In cases where the specification can be met by the inclusion of proprietary equipment, designs or methods of operation, such matters can be discussed with the individual firms concerned in confidence. The confidence is not betrayed by the subsequent enquiry in that the proprietary matters are not proposed in detail. However the final specification can be made sufficiently broad to allow the firms concerned to put forward their methods and still be fully compliant.

It is unlikely, at such an early stage in an enquiry, that the joint meeting will invite collusion between competitors, especially if they are of different nationalities. It also avoids the risk of leakage of estimated prices normally inherent in two-stage bidding.

(c) A procedure suggested by the World Bank in their booklet "Guidelines for Procurement under World Bank Loans". They refer to it as a 'two-step procedure' the first step being an invitation to submit *unpriced* technical bids subject to technical clarifications and adjustments. From this enquiry the preferred technical specification emerges and is issued for the submission of competitive bids. A variation on the theme is to select the preferred contractor on the bids for the first step and to involve him intimately in the later stages of the design development. In such a case the Contract Price has to be negotiated, but if this method of working is foreseen a schedule of rates on which the negotiations can be based can often be included in the first-step offers.

3.4 Final Appraisal and Selection

After all tenders have been received and appraised it will usually be possible and desirable for the Employer (assisted by the Engineer) to reduce the tenders to a short list of not more than about three contenders. It is at this stage that final and detailed negotiations with the preferred tenderer are undertaken aimed at reaching a complete identity of views. In some cases divergences may irrevocably persist and the preferred tenderer be thereby eliminated. The process has then to be repeated with the runner-up. These detailed negotiations are frequently undertaken by the Engineer, especially if the Employer has engaged his consultants in that role. Eventually therefore a recommendation must be made by the Engineer to the Employer and the Contractor appointed. The Employer and the Engineer must keep a constant eye on the date of expiry of validity of the short list tenders and the tempo of negotiations must be adjusted to allow them to be completed before the tenderer becomes entitled to modify his terms and prices.

Chapter 4
Awarding the Contract

The Contract is a legal document binding the parties thereto, i.e., the Contractor and the Employer. It follows that it should be formally awarded and signed by them themselves and not by anybody else professing to represent them. The same should preferably apply also to any "Instruction to Proceed" or binding communication of the same sort: whereas the Engineer may draft them and recommend their issue by the Employer he should not assume he has any implied authority to issue them himself. As far as the tenderer is concerned the Engineer's authority under the Contract does not begin until the Contract is signed, and any earlier powers may merely derive from the Employer having introduced him as "our adviser".

If therefore the Employer requires the Engineer to accept tenders, issue letters of acceptance or instructions to proceed on his behalf, the Engineer must first obtain from the Employer express authority so to act as his agent. He must take steps to protect himself from any liability at law either in respect of payment of the Contract Price or by reason of having misinterpreted the Employer's intentions. The Engineer must always:

- have a written authority from the Employer to act as agent in respect of the Contract.
- have written instructions from the Employer specifying which tender to accept, what negotiated amendments to include, the Contract Price, any temporary limitation on expenditure he must impose etc.
- make it clear in any document he issues that he does so

as agent for the Employer, duly authorised, and himself undertakes no legal liability under the Contract by such action other than that attributed to him as the appointed Engineer under the Contract.

—check the document with the Employer and get his approval before issuing it.

Under no circumstances should the Engineer as appointed Agent execute any deed or bond under seal for his client the Employer. He may do so legally only if he has himself been authorised by a deed—a "power of attorney" and must take special care in such a case not to himself assume unwanted liabilities under the Contract.

Chapter 5
Notifying the Selected Contractor

In cases where it is urgent that work should start on a contract and there is delay in getting the full contract itself or any agreement relating thereto into the Contractor's hands it is usual to give the Contractor advanced notification of the fact that he has been successful, with the aim of getting things on the move.

5.1 Letter of Intent

A letter of intent is the least effective. By its very nature ("We intend placing the contract with you") it cannot itself be 'an unconditional acceptance' and therefore in the U.K. it has no binding effect on either party. Apart from being glad to have the news, the Contractor is fully justified in taking no action and incurring no costs whatever on the strength of it. He has no certainty of recovering the costs if he does. A letter of intent might well be issued by the Engineer on the specific instructions of the Employer in advance of the formal acceptance of a tender by the latter or the drawing up of a formal Agreement for execution by the parties to the Contract.

In some countries overseas, unlike the United Kingdom, a letter of intent can be binding if it fulfils certain requirements. In dealing with overseas tenderers (or when acting for overseas Employers) this point needs to be watched, especially if the laws in the respective countries of domicile of the tenderer and the Employer are different or the contract is to be subject to a legal code native to neither of them.

5.2 Letter of Acceptance

If properly worded by the Employer so as to be 'unconditional' this is a fully binding instrument and sets up a formal contract between the parties. The Contractor is thereby empowered to go full steam ahead on the Contract. It is important, for this reason, that the letter of acceptance should include references to the latest negotiated specification, delivery and completion dates for the Works, the Contract Price, the effective Contract date, the date when the Site is available for Site work to start and any other information of this sort which is essential in any particular contract. Note however that if it introduces any new factor not included in the tender being accepted, it is no longer an 'unconditional acceptance' and at law constitutes a new counter-offer which itself requires unconditional acceptance by the Contractor. Such would be the case if it included any amendments to the tender which had been subsequently negotiated or any changes to the specification for the work. The tenderer's acceptance of the counter-offer must be addressed to the Employer and not to the Engineer for a binding contract to be formed.

An Employer often follows up a valid letter of acceptance with a formal order on his standard order form. In such circumstances it is the letter of acceptance which concludes the Contract and any contract-related dates are referred to the date of acceptance and not the date of the order. The function of the latter is little more than to introduce the Contract into the Employers accounting and administrative routines. Confusion can however, sometimes be introduced if the order is not in identical terms to the Contract or has standard conditions of purchase printed on it and these are at variance with those negotiated in the Contract. If the order is accepted in its turn there is a danger it might be validly regarded as a new contract replacing the earlier one. A note on this subject is given in Chapter 10 below.

5.3 Instruction to Proceed

This is really a limited form of a letter of acceptance and is equally as binding if correctly worded: indeed it should start by a formal statement of unconditional acceptance of the Contractor's tender. It then instructs the Contractor to start work on the Contract or on a specified part of the Contract, and may specify a limit of expenditure the Contractor can incur on the authority of the instruction to proceed alone (e.g. where there are still negotiations in train on some aspects which affect the total Contract Price). The other important points mentioned under Letter of Acceptance above should also be included in an instruction to proceed as may be relevant, and in the same way it requires formal acceptance by the tenderer.

In some quarters, usually connected with the building industry, an instruction to proceed is often loosely referred to as a letter of intent—in view no doubt, of the intention to draw up and execute a formal Agreement. The important distinction, of course, remains whether or not the document itself constitutes an unconditional acceptance of the tender or a part thereof.

Chapter 6
Notes on Standard Forms of Conditions of Contract

6.1 Standard Forms of Conditions of Contract in common use

In this chapter we list the standard forms of conditions of contract more commonly met with in the engineering industry with a few notes on their fields of application. This is not the place to attempt a comparative critique of each, neither does space permit. No two contracts are alike and it is not surprising therefore that the standard forms of conditions as they stand are usually not exactly what is wanted to meet the intentions of contracting parties. Each contract will introduce some necessary variations to a greater or less extent, but the bulk of the chosen standard form of conditions can probably remain unchanged, and their use cuts out the necessity for thinking out and drafting new sets of conditions for each and every contract.

A word of warning on amending standard forms of conditions of contract: as they stand they each form a unified document which has been carefully vetted by legal experts and refined by use over the years. The various clauses are carefully interlinked. Even small amendments need very careful handling; it is alarmingly easy to introduce contradictions, ambiguities and legal loopholes into the structure and such amendments should therefore only be made by those experienced in such matters.

6.2 Standard Forms of Conditions of Contract used by large organisations

In addition to the recognised standard forms of conditions

of contract it is not unusual for large organisations, companies, public bodies and the like to draw up their own standard conditions to suit their own circumstances. They usually expect contractors to accept them in their contracts, albeit in most cases with minor negotiated changes. The extent to which such employers are amenable to amendments varies a great deal. Fortunately in many cases their conditions are derived from and still bear an easily recognised resemblance to one or other of the standard forms of conditions which greatly assists the evaluation of them. In some cases they are in no way one-sided and have been discussed with and approved by the various contractor's associations or organisations whose members are most likely to be affected by them. In others they have been adopted with some minor variations to form what are now equally widely recognised standard forms of conditions. We may, as examples, quote:

- The National Coal Board/BEAMA Standard Conditions of Contract for Plant and Equipment (without erection).

- The British Electricity/BEAMA Conditions (A) (extended Form): Model Conditions of Contract for Plant (including erection). These are based on the IMechE/IEE Model Conditions Form A (see Section 6.4) but have significant modifications.

It is obviously not possible to draw up any comprehensive schedule of organisations which have their own conditions but they include the Central Electricity Generating Board (CEGB), the National Coal Board, ICI Limited, the U.K. Atomic Energy Authority, and The British Steel Corporation to name but a few.

The more common standard forms of conditions of contract met with in the engineering field are listed hereunder. They are all subject to amendment from time to time, clearing up doubts or inconsistencies which have shown up during use or to meet changes in legislation, and on less frequent occasions are revised or rewritten more extensively. When using them in connection with any contract it is therefore necessary in all cases not only to specify the title and version concerned but also the date of the edition, issue or revision and any subsequently issued amendments which are deemed to be incorporated.

Lest this volume should later become misleading, we have omitted in what follows any reference to the edition current at the time of writing.

6.3 Institution of Civil Engineers

"Conditions of Contract and Forms of Tender, Agreement and Bond for use in connection with Works of Civil Engineering Construction". The conditions are widely accepted and used and are generally applicable to all works of civil engineering construction. They can also be used for building works and (rather less satisfactorily and with certain added clauses) for mechanical and electrical works included in contracts largely of a civil engineering or building nature. As they are based on a fully defined specification with bills of quantities, subject to remeasurement, they are not very relevant to contracts for the supply of plant with or without erection on Site. They define considerable responsibilities and duties for the Engineer. They are issued jointly by the Institution of Civil Engineers, the Association of Consulting Engineers and the Federation of Civil Engineering Contractors.

For use with sub-contracts to a main contract which is subject to the ICE Conditions of Contract, there is a printed form of sub-contract ("the Blue Book") the clauses of which are drafted to tie in with the ICE conditions. Blank spaces for details of the two contracts enable the Agreement to be readily completed. The "Blue Book" is published by the Federation of Civil Engineering Contractors.

6.4 IMechE/IEE Model Conditions

"Model Form of General Conditions of Contract recommended by the Institution of Mechanical Engineers, the Institution of Electrical Engineers and the Association of Consulting Engineers".

There are six common versions published jointly by the same bodies:—

- A   Home Contracts—with Erection
- B.1 Export Contracts with Delivery *FOB or *CIF
- B.2 Export Contracts FOB or CIF with Supervision of Erection
- B.3 Export Contracts (including delivery to and erection on site)
- C   Sale of Goods (other than cables) (Home—without erection)
- E   Home Cable Contracts—with Installation.

These conditions are designed and accepted for use with contracts for the supply and erection of almost any form of apparatus, machinery or plant, both at home and abroad. Generally speaking they are unbiassed as between Employer and Contractor and define the duties and responsibilities of the Engineer.

6.5 Institution of Chemical Engineers

"Model Form of Conditions of Contract for Process Plants" These are complete and concise conditions for use with lump sum contracts for supply and erection of process plants in the U.K. In many respects similar to the IMechE/IEE Model Form 'A' they deal in greater detail with matters of commissioning

---

*FOB—Free on Board (i.e. all costs paid across the ships rail at port of loading)
CIF—Cost, Insurance and Freight (i.e. all costs paid unloaded on to quay at port of destination)
For full implications of these and other similar terms see 'INCOTERMS—1953" published by the International Chamber of Commerce, which is the normally recognised authority on the subject in most countries.

and performance of the plant than the Model Form 'A'. They define the duties and responsibilities of the Engineer.

6.6 Standard Form of Building Contract ("RIBA" OR "JCT")

This form is issued by a Joint Contracts Tribunal representing the Royal Institute of British Architects, the National Federation of Building Trades Employers, the Royal Institution of Chartered Surveyors, the County Councils Association and five other associations. It appears in six versions:

- —Local Authorities Edition with Quantities
- —Local Authorities Edition with Approximate Quantities
- —Local Authorities Edition without Quantities
- —Private Edition with Quantities
- —Private Edition with Approximate Quantities.
- —Private Edition without Quantities

The versions "with Quantities" are designed for use in contracts in which accurate bills of quantities (apart from provisional items) can be drawn up at the time the enquiry for tenders is issued. They are not suitable for contracts involving remeasurement (see para. 7.1.5 below), in which cases the versions "with Approximate Quantities" are used.

As might be expected these conditions apply to contracts for the erection of buildings and the like where the Architect has prepared the drawings and acts in relation to the contract in the same way as the 'Engineer' in contracts using the Institution of Civil Engineers or IMechE/IEE Forms of Conditions. They are not usually preferred for building works constructed ancillary to works of civil engineering contracts. The conditions are supported by a series of model forms (such as, for example, "Architects Instructions to Contractors"). These conditions have a number of shortcomings and can be difficult to apply effectively in practice. The supporting documents include three further items of note:

—a comprehensive form of amendments to the "With

Quantities" versions for use when the Works have to be completed in specified sections each by a stated completion date.

—a form of collateral Agreement between the Employer and a nominated sub-contractor dealing chiefly with delay in completion of the sub-contract and the adequacy of any design work involved in the sub-contract (see paragraph 8.3.4).

—a form of sub-contract for use with nominated sub-contractors ("the Green Form"). Although it is published under the auspices of a different authority, this document is intended to conform with the standard forms of building contract. In practice however, main contractors often prefer not to use it when negotiating a nominated sub-contract.

6.7 British Electrical and Allied Manufacturers Association

These are conditions of contract promoted by BEAMA principally for their members' use when quoting for the supply of minor equipment and plant with or without supervison of erection. The wording of the conditions is completely general, however, and their usefulness is in no way restricted to electrical work. They can be used for many of the smaller plant contracts in the engineering field.

There are a number of versions (each a single sheet):—

A	Sale excluding erection—U.K.
AE	Sale excluding erection—Export FOB
AEC	Sale excluding erection—Export CIF
B	Sale including supervision of erection—UK
BE	Sale including supervision of erection—FOB
R	Repair of machinery—UK
RE	Repair of machinery—Export

There are further versions, more specialised and not likely to be of use in general engineering work, so we are not including them here.

6.8 Standard Conditions of Purchase of Goods

Where a contract is to be placed for the supply of goods only, i.e. without erection or supervison of erection, a number of standard forms of conditions of sale (or purchase, as the case may be) exist which are much simpler than those in common engineering use. This is because many of the clauses in the latter relate to the risks and obligations which follow the Contractors need to work on the Employers Site during erection commissioning and testing of the Works.

Some have already been mentioned:—

    IMechE/IEE Model Form C   (Section 6.4 above)

    BEAMA A, AE and AEC   (Section 6.7 above)

Among the others we may mention three by way of example:

- Model Standard Conditions of Contract of the Purchasing Officers Association.

- Standard Conditions of Government Contracts for Stores Purchases Form GC/Stores/1. These consist of Clauses 1-21 which are normally applied to all contracts and 22-59 some of which may be called up by the enquiry.

- The Crown Agents General Conditions of Contract for the Purchase of Goods. These consist of Clauses 1-18 applicable to all contracts and 19-20 called up as required.

For International Use

6.9 FIDIC

These internationally recognised conditions appear in two versions:

— "Conditions of Contract (International) for Works of Civil Engineering Construction"

This has been closely modelled on the Institution of Civil Engineers Conditions (See 6.3 above). A revised edition based on the newer ICE Conditions (Fifth Edition 1973) has now appeared (March 1977).

— "Conditions of Contract (International) for Electrical and Mechanical Works (including erection on site)"

This is the electrical and mechanical counterpart of the foregoing: compare it with the IMechE/IEE Model Conditions Type B.3 (Section 6.4 above) to which it is closely related.

These documents are prepared and issued by the Federation Internationale des Ingénieurs—Conseils (FIDIC) of which the U.K. representative is the Association of Consulting Engineers. The civil engineering version also has the approval of two other international bodies on which the Federations of Building Trades Employers and of Civil Engineering Contractors are represented as well as the Export Group for the Constructional Industries.

6.10 U.N. Economic Commission for Europe

"General Conditions for the Supply and Erection of Plant and Machinery for Import and Export".

These appear in two versions No. 188A and No. 574A which are virtually the same, but 574A has one or two minor changes to meet the needs of contracts with the Comecon countries. Some other contractors might also prefer them!!

There are also two corresponding sets of conditions (Ref. Nos. 188 and 574) for the supply only of plant and machinery.

These four standard forms of conditions are rather general in tone but cover the more important aspects adequately.

6.11 Requirements of Credit-Funding and Insuring Organisations

Organisations concerned with the provision or support of credit facilities for international contracts by U.K. or overseas suppliers usually require, as a condition of their support, that the terms of the ensuing contracts meet with their approval, and under certain circumstances include special clauses which they specify in their regulations. Their object is of course, to safeguard the investments they make and to ensure their funds are not squandered, mis-directed, or placed unnecessarily in jeopardy. We may instance two examples:—

6.11.1 The Export Credits Guarantee Department of the U.K. Department of Trade and Industry (ECGD) is prepared to insure U.K. exporters against failure of an overseas customer to pay the contract price for any of a variety of reasons including insolvency, government intervention, war or insurrection and plain failure to pay.

ECGD requires to inspect and approve the conditions of contract which the U.K. exporter proposes to impose, and to call for changes to any clause they consider inadequately limits their risk. They also have special provisions which they require to be incorporated in certain circumstances, for example those relating to methods of payment from Comecon countries and elsewhere.

6.11.2 The International Bank for Reconstruction and Development ("The World Bank") and the International Development Association which arrange finance for large international projects have comprehensive regulations covering all aspects of contracting to ensure their loans are used for their intended purpose with economy and effectiveness. The World Bank requires to inspect, modify and finally approve:

- all enquiry, tender and draft contract documents
- proposed advertising of the enquiry
- type of contract to be negotiated
- appraisal of tenders received
- negotiations
- final contract documents

to be used by Employers enjoying their loans.

Although the World Bank does not itself publish a standard form of conditions of contract it has definite rules and recommendations as to the type of clause which Employers should incorporate in their proposed conditions of contract when seeking tenders for work to be financed by a World Bank loan.

These requirements are set out in the World Bank booklet "Guidelines for Procurement under World Bank Loans and IDA Credits". The booklet needs careful study as simple sentences, almost casually introduced, often deal with matters of great significance.

Chapter 7
Types of Contract

In this chapter we refer to some of the types of contract that are commonly met with in the engineering field. Each type is classified by a basic feature of the contract, either the way it is negotiated, the manner in which the amount of the contract price is arrived at or some other characteristic. The types are not individually exclusive, i.e. a given contract could involve two or more of the type characteristics and the associated titles.

7.1 Classified by method of evaluating Contract Price

7.1.1 Fixed-price Contract

The Contract Price is fixed and agreed at the time the Contract is signed and does not subsequently change except as a result of variations to the contract specification or under limited circumstances specified in the contract conditions. The Contractor takes all the risk of increased costs due to inflation, introduction of higher taxation, etc., and he will inevitably include in his tender price an allowance for contingencies and for such escalation of costs to cover himself. Thus although the Employer has the advantage of a 'fixed' price, he may, if the Contractor's contingency proves in the event to have been over-generous, pay more than he need have done. If the Contractor's contingency is inadequate he may be tempted to "cut corners" in carrying out the work.

"*Fixed Price*" must not be confused with "*Lump Sum*". "Lump Sum" indicates that the Contractor must execute the contract

Works for an all-in price without any concern by either party as to its sub-division among the component parts of the Works. A "lump-sum" price can be either fixed or subject to price adjustment according to whichever is specified in the Contract. A "fixed-price" contract need not be for a lump-sum: it can comprise any number of separate prices for parts of the Works (e.g. a detailed schedule of prices). Each price is, however, invariable.

Neither must "Fixed Price" be confused with *"Firm Price"*. "Firm Price" refers to a tender price (not a contract price) and implies that the tenderer is prepared to be held to the price and enter a contract on it. The opposites of "firm price" are "approximate price", "estimated price", "budget-price" or even the unusual (American) "ball-park figure". A "firm price" subsequently embodied into a contract does not necessarily become a "fixed price": it can still be subject to cost price adjustment if the contract conditions so provide.

7.1.2 Price-adjustment Contract

This is a contract which includes a price adjustment clause (CPA clause–'Contract Price Adjustment') whereby the contract price is increased or decreased as a result of changes up or down of certain specified costs which the Contractor incurs. Under conditions of inflationary uncertainty a CPA clause is almost essential in any contract which is likely to take more than 12 months to complete, for the Employer's benefit as well as the Contractor's. The latter is protected (in good measure if not completely) against unpredictable cost increases; the Employer besides avoiding excessive contingencies in the contract price does not have to deal with a Contractor trying to recoup a loss position by skimping work, multiplying claims and, in the extreme case, appealing for ex gratia help to keep going on the work.

The extent of the adjustment permissible is usually specified by a formula in the CPA clause agreed between the parties and is most conveniently based on some recognised statistics or indices of ruling costs such as those published from time to time by government departments, industrial associations or suchlike for such a purpose. This safeguards the interests of both parties.

Some conditions of contract, however, do not tie the permissible price adjustment to a formula but require the Contractor to justify any claim for an increase. One such example is the I.Mech.E./I.E.E. Model Conditions referred to in Section 6.4. This justification can be laborious and very time consuming especially with contracts not based on a detailed schedule of rates. It involves maintaining full records not only to establish the rates and prices on which the tender price was based, but to show that increases in cost have, in fact, been experienced, their amount and (often most complex of all) the point of incidence in time when increases reflected on to the contract costs. Increases in overheads are difficult to substantiate as are labour costs by any contractor who habitually pays his employees more than the standard nationally-agreed rates. Potential sources of disagreement between the Contractor and the Engineer are many and it is therefore important whenever possible to negotiate the inclusion of a suitable formula and sources of indices into such conditions of contract.

Formulae almost invariably take the general form:

$$P_R = P_B \left[A + B \times \frac{L_R}{L_B} + C \times \frac{M_R}{M_B} + D \times \frac{N_R}{N_B} - - - \right]$$

where:

P_R = Adjusted Contract Price

P_B = Original Contract Price

A,B,C,D etc. = The fractions of the Contract Price repre- by different services or commodities such as labour, transport, steel, copper, oil, etc. (The total $A + B + C$ etc., must, of course, total 1.00)

A clearly represents a fixed part of the Contract Price.

L_R, M_R, N_R etc. = The respective indices applicable at the date of revision.

L_B, M_B, N_B etc. = The respective indices at the base date (used by the Contractor in calculating his tender price).

Formulae commonly used in the civil engineering/building fields are those derived from an original provided by the National Economic Development Office (NEDO), for which specially loaded indices are published monthly by the Property Services Agency of the Department of the Environment. That developed for the civil engineering industry (usually known as "the Baxter Formula") is based on a fixed fraction and eleven variable fractions which are agreed between the parties to the contract or estimated by the Engineer for each contract. They cover labour, provision and use of contractors equipment and transport, and nine groups of materials. There is a special version applicable to structural steelwork in which the method of application of the formula is modified to take account of fabrication off-site.

A second NEDO price adjustment formula is designed for use with building works (often known as "the Osborne Formula"). By virtue of some 48 basic work categories and 5 others for specialist engineering installations (Electrical, Heating and Ventilating, Lifts, Structural steel and Catering Equipment) the Osborne formula is considerably more complicated in application than the Baxter formula. The indices (including those for the specialist engineering installations) are also published monthly by the Property Services Agency referred to above.

In another formula used for contracts involving mechanical or electrical plant (the BEAMA formula) the number of terms is reduced to two—one for labour and one for materials. It is published by the British Electrical and Allied Manufacturers Association and uses as indices for materials those published monthly in the DTI Trade and Industry Journal—Table 1 either for Electrical Machinery or for Mechanical Engineering Materials. For labour it uses figures issued monthly by BEAMA based on

National Average Earnings Indices for the Engineering Industry produced by the Department of Employment. The BEAMA formula is in no way tied to the BEAMA standard forms of conditions or to purely electrical apparatus: it can be used with a wide range of contracts for the supply of plant of a mechanical or electrical (or mixed) nature, with or without erection. BEAMA also publish a number of variants of their main formula using factors and constants appropriate to different types of project, e.g. turbo-generation plant or electronic equipment, and for export purposes.

A closely parallel set of formulae useful in contracts dealing with pipe work and similar steel-based products are the three versions published by the Water-Tube Boilermakers Association. These again employ only two variable factors, materials (usually based on the DTI monthly figures from Index Nos. 311 and 312 (table 2) from the T and I Journal mentioned above) and labour (based on WTBA's own index which in its turn is based on the labour rates and allowances nationally accepted by the Engineering Employers Federation). The three versions referred to have different values of A, B, C (in the formula above) designed to apply to home contracts including erection, home contracts excluding erection, and overseas contracts respectively.

Some formulae (including the BEAMA) use as the 'factor applicable at the date of revision' an average of the monthly indices over a part of the contract period. This allows, for example, for the fact that in plant manufacture, materials are all purchased early in the contract period whilst labour is mostly involved in the later stages of the contract period.

Similar formulae are associated with a number of other standard forms of conditions of contract such as, for example, the United Nations Economic Commission for Europe—188, and many firms have their own versions in which the components of the formula (L,M,N etc.) are specially selected to fit their own products. A transformer manufacturer, for example, might well select labour, transformer steel, copper and oil.

7.1.3 Cost-plus Contract

This is a contract in which the Contractor is reimbursed his actual costs incurred, plus specified percentages on the actual costs in respect of his overhead charges and his gross profit, without any total contract price being quoted. Its main use is in cases in which the scope and progress of the Contract cannot be foreseen (as for example in contracts for research and development or in areas of limited knowledge). Many of the early contracts in the atomic energy and space fields had to be of this type.

Cost-plus Contracts all have the basic disadvantage from the Employer's point of view, that it is not necessarily in the Contractor's interest to minimise his costs and there can therefore be a conflict of interest between the Employer and the Contractor. An increased burden of close control is placed on the Engineer. Cost-plus contracts need very careful negotiation and drafting to safeguard the Employer's interests by suitable constraints on the Contractor but the danger is never fully removed: constant and detailed supervision of the progress of the work (both technically and administratively) forms the best safeguard. Clearly the Contractor must make all his records and accounts available for inspection by the Employer or by some agreed neutral third party such as a firm of chartered accountants, a quantity surveyor or the Engineer, and there must be clear and detailed agreement from the outset on just what the Contractor can legitimately include as his costs and expenses and what rates of pay, etc., he can make. Rates of pay may become a very thorny point in cost-plus contracts involving specialists who are in short supply as there is no restriction on the Contractor offering unnecessarily glamorous wages. The cost of making good faulty work is another difficult area. Since by definition it is impossible to calculate a realistic cost for a 'cost-plus' project, competitive negotiation of a contract will of necessity be based on technical aspects alone, i.e. by discussing the project with possible contractors and deciding which appears to be technically the most competent, most suitably orientated and most efficient (and thereby most likely to incur the least costs). Efficiency must have regard also to such matters as labour relations and work control.

7.1.4 Target-Cost Contract

This is an elaboration of the 'cost-plus' concept which aims to overcome the main defect of the latter by introducing incentives to the Contractor to be as efficient as possible. Its main use is in large contracts in which the urgency calls for tendering to go ahead before the detailed design has been done. It presupposes that accumulated experience will enable a figure to be set with some accuracy for the total cost, even though the breakdown by quantities may not be possible at the time of tendering. An example might be the structural work for a modern power station. In its simplest form a target-cost for the contract is established and agreed between the Employer and the tenderer aided by the Engineer as impartial mediator. The Contractor is eventually paid cost-plus against his actual expenditure or on the basis of remeasurement. If his total costs prove to be less than the target-cost he gets as a bonus profit an agreed proportion of the difference; if his actual costs exceed the target-cost he gets paid only an agreed proportion of the excess, the Employer retaining the rest as his bonus. Such proportion can be on a sliding scale.

The importance (and difficulty) of fixing the target-cost with some accuracy is clear: if it is fixed too low there is no incentive for a tenderer to undertake such a contract with all its attendant risks and accounting troubles and he will stick out for a normal cost-plus contract with its absence of restraints. If it is fixed too high it is too easy for the Contractor to make a large profit at the expense of the Employer. Negotiating the target-cost can easily become something of a poker game between the Employer and tenderer, and hence the need for past experience of similar projects. The size of the project must be sufficiently large or diversified that if some of the quantities prove to be abnormal, their effect on the total is still negligible; or failing that, a situation of "swings and roundabouts" will help to neutralise miscalculations overall.

More elaborate versions of target-cost contracts have been

devised to allow of competitive tendering in which each tenderer establishes his own target-cost and percentage return. They introduce many problems, both of estimation and appraisal but are useful in cases in which target-cost principles have to be applied to contracts involving public funds. In such cases competitive tendering is normally essential so that simple cost-plus contracts are unacceptable, except as a last resort.

7.1.5 Bills of Quantity Contract

This is a common form of contract in engineering (especially civil engineering and building projects) in which the design has been completed by the Employer and can be specified in the tender. The Works are broken down into as many individual parts or activities as are deemed necessary and these are billed in the tender enquiry together with the quantity of each item involved. The quantities are obtained by measurement from the project drawings or by other methods and each tenderer is required to price all the items in detail. The contract conditions must specify whether these measured quantities are firm and fixed or whether they are subject to remeasurement, i.e. to redetermination on completion of the actual work on Site. Such remeasurement is the responsibility of the Engineer and is standard procedure on civil engineering and building contracts. The Contract Price is determined by applying the billed rates to the remeasured quantities.

The establishment of bills of quantities and the procedure for measurement and remeasurement in connection therewith are largely standardised and regulated in the engineering and building industries by two widely recognised documents:

- In civil engineering: "Standard Method of Measurement of Civil Engineering Quantities" published by the Institution of Civil Engineers.

- In the building and constructional field: "Standard Method of Measurement of Building Works" published jointly by the Royal Institution of Chartered Surveyors and the National Federation of Building Trades Employers.

Bills of quantity contracts offer several advantages to the Employer, notably:

- Each tenderer establishes his prices on the same basis and their detailed rates and prices can be compared.
- Rates for each item are automatically established and can be used where applicable to ascertain the value of variations to the Works.
- If the bills are drawn up competently for the Employer, it ensures that the tenderer (who may not have much time to absorb and establish the intricacies of the project) does not omit any significant costs from his tender price.

Before the tender enquiry is issued all provisional items and those which are the subject of prime cost sums are included in the bills and in each case the Employer inserts the amounts he wishes to be included in the contract for them. These sums together with the totals of the bills after pricing by the tenderer give together the total tender price.

7.1.6. Schedule of Rates Contract

This is an alternative to the bills of quantities contract used in cases in which for any reason the quantities of the items of the bills of quantities cannot be established with any reasonable accuracy at the tender enquiry date. Broad estimates are assessed (possibly with upper and lower probable limits) and the tenderer is required to quote a schedule of rates only, based on the type of work and general size of the different items. Payment to the Contractor under a schedule of rates contract is made on measurement of the work eventually done. The measurement is the responsibility of the Engineer on behalf of the Employer, or alternatively measurement by an appointed firm of quantity surveyors may be accepted by the Engineer.

A schedule of rates contract might for example be used in cases in which it is desired to place a contract before the specification has been finalised and is liable to considerable alteration, e.g. in a negotiated contract (see 7.2.2 below). It is often used in the petrochemical industry. It is also commonly used for installation works for mechanical or electrical services (e.g. cabling or piping) when exact routes are not predictable.

7.2 Classified other than by method of evaluating Contract Prices

7.2.1 Competitive Contract

A Contract arrived at by the process of formal competitive tendering by a number of tenderers against a common specification.

7.2.2 Negotiated Contract

A Contract negotiated between the Employer and a potential contractor of his choice. It is sometimes adopted when the project specification is not clear cut, but is arrived at by discussion on ways and means between the Employer and a specialist contractor. Such a situation may often apply, for example, to the cost-plus type of contract (see 7.1.3 above), or for specialised plant or services in which there is only a single supplier in the field, or to nominated sub-contracts.

If the Employer fails to negotiate an agreement with his selected contractor he is at liberty to try again elsewhere, but in doing so he must be very careful how far he makes use of information or know-how obtained from the first contractor during the earlier negotiations: a clear understanding at all stages on what is confidential and what is not must be established. Experience shows that a negotiated contract can take just as long to finalise as one by competitive tender and it should not be assumed that time will be saved. It is not unknown for the contractor to strengthen his monopoly position by delaying progress until it is too late for the Employer to take alternative action. Whenever possible, therefore, an Employer attempting a negotiated contract should do so strictly to a timetable which enables him to go to tender in a different way if negotiations become protracted. Earlier in this volume (Section 3.3) we described a hybrid form of contract—part negotiated, part competitive—

which can sometimes be used to overcome the 'monopoly' aspect of a pure negotiated contract. It should be referred to at this point.

It must be appreciated that in any sizeable engineering contract which has been out to competitive tender there will usually be some points of difference between the Employer and the selected tenderer which are then removed by negotiation. The term 'negotiated tender' or 'negotiated contract' is not usually applied to such negotiations unless the areas of divergence are so wide that the competitive tender played but a small role in arriving at the contract.

7.2.3 Package Contract (Package Deal)

This is a contract in which two or more related jobs, each of which could form a separate contract, are combined and placed as a single contract. In the engineering field it often involves combining design and/or development with construction or supply and erection and/or maintenance. By reason of the incomplete specification which is unavoidable when design or development is embraced by the package contract, it is usually necessary to allow flexibility in the Contract Price such as by the use of a 'cost-plus' or 'schedule of rates' form of contract.

The Contract must clearly lay down who is responsible for selecting the actual development design which is to be used for construction or manufacture so that if the first embodiment fails on completion to meet the functional specification there will be no doubt who pays for the necessary redesign/modification of the works. In the civil construction and building fields a package contract usually requires the Employer (through the Engineer) to approve and accept the design before construction starts and thereafter the Employer is responsible for any defects in the design. It is often referred to as a 'design and construct' contract. On the other hand in the mechanical or electrical engineering fields the design usually depends on the expertise of the Contractor who will also manufacture the plant and the responsibility for the correctness of the design therefore usually remains with the Contractor. Any approval of the design or the drawings by the Engineer prior to manufacture is normally to ensure the proposals are generally in line with the Employer's needs and in a suitable state to allow expenditure on manufacture to be incurred: no technical approval or liability is thereby implied. This must be made clear in the contract documents. As in any contract involving design or development work, the respective rights of the two parties in the resulting design must be specified and the conditions under which (if at all) the Contractor may employ such design for his other customers, or the Employer for further repeat projects, must also be established in the contract documents.

The precautions which the Employer (and in practice, the Engineer on his behalf) must take to protect the Employer's interests in various types of contract with flexible contract prices have been referred to under their respective headings above: in the case of a package contract with a design element he must also constantly appraise the technical development to avoid the adoption of an unnecessarily costly or elaborate solution which would involve unnecessary expense at the constructional stage.

The advantage of a package contract to the Employer is continuity of technical and administrative responsibility in what can often prove to be a difficult sequence, more especially in 'design and construct' contracts. Other advantages may be a reduction of the number of contractors employed on Site or, on occasion, an inducement to a contractor to undertake an unattractive contract by joining it with an attractive one.

7.2.4 Turnkey Contract

A turnkey contract is a package contract in which all the separate disciplines, civil, mechanical and electrical for the setting up of a major entity are placed in the hands of a main contractor, who may sub-contract specialist works to appropriate sub-contractors. The advantage to the Employer is that having specified the major entity and selected the main contractor he is himself relieved of all detail and the co-ordination of the work of sub-contractors and the interworking of their separate undertakings. All of these are carried out by the main contractor, the Employer's interests being watched by the Engineer throughout. The Employer expects to take over the Works as a fully operational complex of proven performance. As all three disciplines are involved, the conditions of contract for turnkey contracts frequently present problems of great complexity. Standard forms of conditions of contract are often quite inadequate and need carefully worded additional clauses and amendments.

7.2.5 Continuation Contract

A continuation contract is a contract negotiated with a Contractor already at work on an existing contract with the Employer, and based on the existing terms and conditions. It gives both parties continuity of action between firms who know each other and presumably have found they can work together satisfactorily. It frequently permits the pooling or rapid switching of plant, machinery, and specialist labour between two similar projects with consequent savings in time and money. A continuation contract may be for a quite different project from the one on which the Contractor is currently working. If it does have a connection it is sometimes referred to as an *Extension Contract*.

Price negotiations for continuation contracts may not be easy as the Contractor is well able to appreciate the strength of his position and the advantages the Employer stands to gain.

7.2.6 Serial Contract

This is a similar arrangement to a serial tender (see Section 9.4) but with a firmer commitment based on a contract, which not only deals with the first project but makes provision for a series of specified additional projects to be called up later.

7.2.7 Running Contract

A contract to provide goods or services at specified intervals or as required from time to time by the Employer over a stated period of time (often 1-2 years). The Contractor is usually the exclusive supplier during the period and his prices are quoted against an estimate of total demand and probably a guaranteed minimum value but no total figure is fixed. Price rates may be fixed for the period or a contract price adjustment arrangement may be included.

7.2.8 Service Contract

A service contract is one concerned solely with the provision of services in contra-distinction to the carrying out of work or supplying goods or equipment (with or without installation). Thus contracts for drawing and design, research and advice or the maintenance of plant after completion are examples of service contracts. A consultant's contract with his client is a service contract. The term is also used for a contract with a public utility company to provide a "service" of electricity, gas, water or the like.

Chapter 8
Assignment and sub-letting of Contracts

8.1 Assignment

Standard forms of conditions of contract (and others) usually contain a clause prohibiting the Contractor from "assigning the Contract or any part thereof or any benefits or obligations therein without the written consent of the Employer". Its implications may be somewhat obscure: the transfer of parts of the works by sub-contracting is a recognised procedure. Furthermore the Contractor has a right to receive payment for the Contract and to use the money once received in any way he wishes. Also the law does not permit unilateral transfer of liabilities so that a Contractor could not escape his contractual responsibility for completing the work by a purported 'assignment' (legally—except on death or bankruptcy—obligations cannot be assigned, only rights or benefits).

Such a clause must therefore usually be interpreted more narrowly as preventing (without the written consent of the Employer) the transfer by the Contractor to another party of the whole contract works and especially his basic functions of administration, organisation, accounting and control of the contract and the co-ordination of the sub-contractors. The clause also prevents the Contractor assigning to a third party the right to be paid part or all of the Contract Price direct by the Employer (with an implied right to sue the Employer for non-payment).

8.2 Sub-letting and Sub-contracting

8.2.1 In parallel with assignment most standard forms of contract conditions have a clause restricting the Contractor from sub-letting (sub-contracting) any part of the works without the written consent of the Engineer. The clause usually goes on to state that the Engineer's approval and consent so given in no way relieves the Contractor of his responsibilities for the work sub-let but that he shall remain totally liable to the Employer for the "acts defaults and neglect" of his sub-contractors.

8.2.2 Certain sub-contractors may be specified by the Engineer in the tender enquiry (for example, for the supply of stated proprietary machinery or the erection of structural steel) and in addition the tenderer may be required to nominate in his tender the firms to whom he proposes to sub-let defined parts of the Contract. In both cases the Engineer's approval or otherwise will be inherent in any contract eventually negotiated and placed. There would still remain for separate submission to the Engineer on the Contractor's initiative, any sub-contracts proposed after the date of the Contract or introduced by subsequent variations to the specification for the Works.

8.2.3 Conditions of contract rarely indicate any criteria on which the Engineer should decide to approve or reject a proposed sub-contractor. He must consider inter alia:

- the sub-contractor's technical suitability for the tasks it is proposed to allot him
- his financial stability, existing commitments and further creditworthiness
- his reputation for prompt and reliable work of high quality, and good worker-relations.
- any 'political' or security implications from the standpoint of the Employer, the Engineer, the Contractor.
- any preferable alternative sub-contractor—on grounds of ability, experience, reliability, location, price, or the like.

8.2.4 It is frequently provided that "the Engineer's approval shall not be unreasonably withheld". The Contractor has been made responsible for producing a result and must be allowed to set about the task in the way he has decided is best, unless the Engineer has very strong reasons for requiring him to change. However, in considering approval the Engineer must always have in mind that an Employer normally has no legal relationship with a sub-contractor. He has no direct recourse against him for faulty work, neglect or other breach of his sub-contract. The Employer can only hold the Contractor (with whom he has a contract) responsible and hope that the terms of the sub-contract are such as to enable the Contractor to take what action is necessary to get the matter put right. We deal with this point further in Section 8.3 below in relation to the special case of nominated sub-contractors.

8.3 Nominated Sub-Contractors

8.3.1 It is generally fair to say that the nominated sub-contractor concept is peculiar to construction contracts where the Employer/Engineer customarily specifies the works to be constructed in full detail. Contracts for supply of mechanical or electrical equipment are usually based on a functional specification leaving design detail and provisioning matters largely in the Contractor's hands. The use of a proprietary article may sometimes be specified in such contracts but this usually only adds one more item to a list of purpose-built components (e.g. castings, cabinets, geartrains, control equipment) already used by the manufacturer and no new problem is posed.

8.3.2 It is commonly supposed that the main incentive for the Employer to nominate work as sub-contracts springs from his desire to have specialist work carried out by specialists he considers particularly competent. On the contrary the main incentive is cost.

Work within a contract which is beyond the capability of the ordinary construction contractor will invariably be sub-let. This means that all the tendering main contractors would need to make many enquiries of specialists in order to secure a keen price. Usually this will need to be achieved by obtaining competitive sub-tenders often involving much duplication of effort. The process thus leads to a likely increase in the time and cost for the submission of main tenders.

The Employer, by removing the work from the main tenderers area of pricing, retains control over the right to select and therefore retains price control for the sub-contracted work which he would lose if the right to select remained with the main contractor.

8.3.3 Of recent years many contract draughtsmen have assumed that control over the quality of the sub-contractor rather than price is the principal object of the system. In most cases the Employer's view would surely be that control over quality is but a necessary requirement in order to achieve the principal object, which is control over price.

It is most important to recognise this motivation, for misunderstanding has conceivably led to the present situation whereby the Contractors responsibility for all his sub-contractors to the Employer is seriously impaired. So much so that any advantage that the Employer might gain on cost is seriously imperilled by increased risk.

Anybody wishing to be convinced might do well to study a critique of the ICE 5th Edition Standard Form of conditions in respect of nominated sub-contracts by Mr. Max Abrahamson, a well-known contracts lawyer, which was published in the journal 'New Civil Engineer' for 13th March 1975, pp. 43-46.

Legal actions have been plentiful, the main results of the Court decisions being to high-light new dangers. Mostly the problems react to the disadvantage of the Employer and the inevitable conclusion is that whenever possible nominated sub-contracts must now be avoided.

8.3.4 The dangers to the Employer arise mainly from the absence of any contractual relationship between him and the subcontractor as mentioned earlier: he can neither influence the performance of the sub-contract direct nor sue the sub-contractor if he becomes in breach of his sub-contract. The only recourse is second-hand, through the main Contractor who had no part in selecting the sub-contractor (indeed he may have been actively opposed to the idea) and who is much more concerned with his own difficulties in the situation rather than with the problems of the Employer.

Probably the most common risk areas are:

—when the nominated sub-contractor is in delay on completion, in spite of the best efforts of the main Contractor
—when design work for which the nominated sub-contractor is responsible (e.g. for plant supplied) is inadequate or unsuitable or unsafe and the main Contractor has no design responsibility to the Employer in his main Contract.

The situation is sometimes met by the Employer entering a collateral agreement direct with the nominated sub-contractor as a prerequisite to nominating him. The agreement is designed to make the sub-contractor directly responsible to the Employer for prompt performance and acceptable design in consideration of which the Employer undertakes to nominate him to the main Contractor. The Standard Form of Building Contract (see Section 6.6 above) typically illustrates the difficulties described above. The ancillary documentation issued by the Joint Contracts Tribunal for use with the Standard Form includes a recommended form of agreement between Employer and nominated sub-contractor such as we have described.

8.3.5 With construction contracts based on the ICE 5th Edition Standard Form of Conditions or the Standard Form of Building Contract, nominated sub-contracts are included whenever a prime cost sum is included in the bills. They may also arise by the Employer/Engineers instructions given in regard to the expenditure of provisional sums included in the bills.

8.3.6 The term 'nominated sub-contract' is not confined to the two sets of conditions mentioned in 8.3.4 above but is used elsewhere in other standard conditions of contract. The meaning ascribed to the phrase will depend on the contract definition provided. Similarly, the terms 'prime cost sum' and 'provisional sum' have no meaning and should therefore be defined in the Contract, though it may be noted that no such definition is provided in the Standard Form of Building Contract.

Chapter 9
Types of Tender

Although an exposition of tenders and tendering procedure is outside the scope of this present volume, it may be useful to include here a note on the more important types of tender, especially as they tie in with the types of contract.

9.1 Open Tender

This is an invitation to tender at large by means of notices placed in the national, international or technical press. It is little used in the engineering field except by public authorities, nationalised industries and the like or for the supply in bulk of everyday materials such as cement.

Under EEC Rules, large public works contracts (i.e. building and civil engineering) being offered by the following authorities must be advertised in the official EEC Journal so that contractors in all EEC countries have an equal opportunity to tender. The authorities are:

- local authorities
- new town corporations
- Commission for the New Towns
- Scottish Special Housing Association
- Northern Ireland Housing Executive.

Nationalised industries and government controlled organisations (e.g. BBC, Electricity Authority, etc.) are all exempt.

"Public Works" excludes:

- Industrial plants of mechanical, electrical or power nature (production and/or distribution)
- Nuclear plants (industrial or scientific)
- Works specifically in connection with mineral extraction (e.g. drifts, shafts, tunnelling, etc.)
- Variations to existing contracts and repeat contracts
- A number of specialist exclusions.

Full details can be obtained from DOE circulars 4/73 and 59/73 which also give reference to the applicable EEC directives.

9.2 Selective Tender

Invitation to tender sent to a selected list of tenderers. The list is usually arrived at by one of two procedures:

- Invited list procedure. The list is drawn up ad hoc with the particular contract in mind.
- Standing list procedure. A standing list of approved firms is maintained from which a short list is drawn for each contract.

In either case a preliminary enquiry is usually sent out to eliminate those contractors who for various reasons do not wish to tender for that particular enquiry. The enquiry documents for large contracts are both bulky and expensive: they cannot be wasted on non-starters.

9.3 Negotiated Tender

The project is discussed and negotiated with one preferred firm which, when the details of specification and contract have been settled, makes a tender offer, also open to negotiation. The preferred firm may be selected on past experience, technical, topographical or similar grounds or by calling for estimated prices based on outline contract terms from a number of firms but this last course can only be adopted if a reasonably clear specification can be given to them. In suitable cases the tender offer may be replaced by a "cost-plus" or target-cost agreement.

9.4 Serial Tender

A serial tender is similar in intent to a serial contract, but the latter more firmly commits the parties to the later projects after the first. The serial tender is a standing offer whereby a contractor undertakes to enter into a series of separate contracts spread over a stated period in accordance with the terms and conditions of the standing offer. Sometimes used for supply of requirements of materials (e.g. cement) at intervals as required over a period of say two years. The standing offer usually includes an intention to supply goods or to do work to a stated minimum value (subject to satisfactory performance by the Contractor) and may in addition indicate without further commitment the anticipated total value. It enables the Employer to have virtual 'call-off' facilities without having to negotiate a contract each time and without having to decide on a firm total requirement some time in advance. He can expect to get a better 'bulk' price than he would on each small individual contract as the Contractor can plan the whole programme ahead.

9.5 Requirements of the World Bank and I.D.A.

We have already referred in paragraph 6.11.2 to the regulations which the World Bank imposes on Employers using the proceeds of its loans when entering into contracts with suppliers, and we indicated how they laid down the type of tender ("form of procurement") and tendering procedure to be adopted in some detail.

For large or complex contracts the World Bank requires the issue of an enquiry for prequalification bids which has to be notified and advertised by the same methods and in the same countries as their regulations lay down for tender bids. The Prequalification enquiry sets out the scope of the Works, an abbreviated specification and a clear statement of the requirements for qualification in a manner to elicit from the bidder his ability to execute such a contract, under three main headings:

—experience, and past performance on similar contracts
—availability of adequate and suitable personnel, equipment and plant
—financial status.

As with other documentation the Bank requires the Employers proposals to be submitted to them for comments and approval before they are issued to tenderers.

Chapter 10

Conditions of Contract printed on order forms

In this volume we have had in view, almost exclusively, contracts of the type with which consultants are largely concerned, namely the large type of contract negotiated formally and documented individually. It might nevertheless be of interest in conclusion to illustrate the interplay of offer and acceptance in forming a contract by considering the standard conditions of purchase which so many companies print, usually in very small type, on the back of their order forms. Usually these conditions are introduced by a statement (also in small type) somewhere on the face of the form that "This order is placed subject to our standard conditions overleaf". They provide a good example of the way these matters can sometimes go adrift.

What happens? Firm "A" (the Employer) issues an enquiry ("Can you supply me with . . . .?") without specifying any conditions of contract. Firm "B" tenders subject to its chosen conditions of sale. Firm "A" places an order on its normal order form as above described and even specifies its purchase as "Generally in accordance with your tender dated so-and-so". The order form has a tear-off acknowledgement and acceptance slip which the Contractor, Firm "B", is requested to sign and return. This he dutifully does under the comfortable impression that his selected conditions of sale (part of his tender) have been established in the Contract.

But have they? What is the real position? Employer "A" by the conditions of purchase called up by the printing on his form, and by using the unspecific phrase "generally in accordance with your tender" has not accepted Contractor "B"s tender unconditionally but has made a counter-offer. Contractor "B" by completing the acceptance slip has given his unconditional acceptance thereof. He has himself a contract—but subject only to the Employer's conditions of purchase as printed on his order form and not the Contractor's own conditions of sale.

This situation arises continually and needs constant vigilance. Even if the order were placed "Fully in accordance with your tender . . . . . .", the Contractor would be well advised to seek confirmation that the printed references to the Employer's conditions of purchase are withdrawn. The order itself then constitutes an unconditional acceptance of the Contractor's tender and the Contract is made accordingly.

Appendix 1

EQUIVALENT TERMS USED IN SOME COMMON STANDARD FORMS OF CONDITIONS OF CONTRACT

The terms shown in any line have generally the same meaning but there are cases where the definitions and use are not identical. The terms for FIDIC (Civil) and FIDIC (Electrical & Mechanical) Conditions are the same as those used in the ICE (5th Edition) and IMechE/IEE Model Form "A" Conditions respectively.

I.C.E. (5th Edition)	IMechE/IEE "A"	IChemE	Standard Form of Building ("RIBA" or "JCT")	EEC (574A)	BEAMA "B"	This Volume
Employer	Purchaser	Purchaser	Employer	Purchaser	We	Employer
Contractor	Contractor	Contractor	Contractor	Contractor	You	Contractor
Engineer	Engineer	Engineer	Architect/Supervising Officer	–	–	Engineer
(Temporary) Works (Permanent) Works	Works	Works	Works	Works	–	Works
–	Plant	Plant	"materials and goods"	Plant	Goods	Plant
Construction Plant	Contractor's Equipment	Contractor's Equipment	–	Contractor's Equipment	–	Contractor's Equipment
Completion	Completion	Completion	Practical Completion	Completion	Completion	Completion
⎱ ⎰	Taking Over	Taking Over	"taking possession of"	Taking Over	Taking Over	Taking Over
"final test"	Tests on Completion	Take-Over Tests	–	Take-Over Tests	–	–
Cert. of Completion	Taking Over Cert.	(Taking Over Cert.) Acceptance Certificate	Certificate of Practical Completion	Take Over Cert.	–	–
"a certificate"	Interim Certificate	"a certificate"	Interim Certificate	–	–	–
Maintenance Certificate	Final Certificate	Final Certificate	Final Certificate	–	–	–
Period of Maintenance	(no specific term)	Defects Liability Period	Defects Liability Period	Guarantee Period	–	Period of Maintenance
Contract Price	Contract Price	Contract Price	Contract Sum	–	Contract Price	Contract Price
–	Contract Value	–	–	–	Contract Value	Contract Value

97

Index

(References are to the numbered sections of the work.)

A

Acceptance, Letter of	4, 5.2
Acceptance, necessary for contract	1.1
Action, limitation on period before starting	2.13.1
Agent, Engineer as Employer's	2.14.3, 4
Agreement, collateral, with sub-contractor	8.3.4
Agreement, Form of	2.12, 6.6, 8.3.4
Agreement, formal	1.3, 2.10.1, 2.11, 2.12
Agreement necessary for contract	1.1, 1.2
Appraisal of tenders	3.4
Assignment of rights	8.1

B

Bank, The World	3.3.2, 6.11.2
Baxter formula for CPA	7.1.2
BEAMA formula for CPA	7.1.2
Bills of Quantities	2.6, 6.3
Bills of Quantity Contracts	7.1.5
Bonds	1.7, 2.11, 4
Bonds, tender	2.11.2, 2.11.3
Bonds, performance	2.11.2
Bonds, repayment	2.11.2
Bonds, plant performance	2.11.2
Breach of contract	1.9

C

Capacity of parties to contract	1.5, 2.13, 4
Competitive Contract	7.2.1
Consideration of value to pass	1.7, 2.13.1, 2.14.2
Continuation contract	7.2.5
Contract, breach of	1.9
Contract, conditions of	1.9, 2.8, 7.2.4, 10
Contract conditions, legal necessity for	2.8.4
Contract conditions, modifications to	2.8.3, 2.8.5., 3.2, 6.1, 6.2
Contract conditions, special	2.8.2
Contract conditions specifying Agreement to be signed	2.12
Contract conditions, standard forms of	2.8, 6
Contracts with Comecon countries	6.10, 6.11.1
Contract documents	Def. 2.2, 2.11.3
Contract evidenced in writing	1.8
Contract by exchange of letter	1.8.1, 10
Contract involving design and development	2.6.2, 7.1.3, 7.2.2, 7.2.3, 8.3.4
Contract in writing	1.8
Contract, objective in draughting	2.1

99

Contract price	Def. 2.6.1, 2.10.2, 2.11.2, 2.14.1, 7.1.2, 8
Contract, right to determine	1.4, 1.9, 2.7.1, 2.11.2
Contract, variations to specification	2.6.1, 2.6.3, 2.14
Contract, variations to terms and conditions	2.14
Contract value	Def. 2.6.1
Contract, verbal	1.8
Cost-plus contracts	3.3.1, 7.1.3, 7.2.3, 9.3
Cost price adjustment (CPA)	7.1.1, 7.1.2

D

Damages	1.4, 1.9, 2.7.1, 2.11.2
Damages, liquidated	1.9, 2.7.1
Data, Site	2.4
Deeds	1.7, 2.11, 4
Delay in Contract performance	1.9, 2.7
Documents, contract	Def 2.2, 2.11.3
Documents under seal	2.11.1, 2.13
Duress	1.4(c)

E

Engineer, powers of	Def. 2.14.3, 4, 6.3, 6.4, 6.5
Engineer, role of	Def. 4, 5.1, 7.1.3, 7.1.5
Evidenced in writing, Contracts	1.8, 2.11.1
Export Credits Guarantee Dept (ECGD)	6.11.1
Extension Contract	7.2.5
EEC Rules	9.1

F

Firm-price tender	7.1.1
Fixed-price contract	2.6.1, 7.1.1

G

Guarantees	1.7, 2.11

H

Hand, Agreements under	2.13

I

Influence, undue: on party to contract	1.4(d)
Instruction to Proceed	4, 5.3
Intent, Letter of	5.1, 5.3
International Development Association (IDA)	3.3.2, 6.11.2

L

Legal relationship necessary for contract	1.3
Legality of objects of contract	1.6
Letter of Acceptance	4, 5.2, 5.3
Letter of Intent	5.1, 5.3
Letters to form contract, exchange of	1.8
Liquidated damages	1.9, 2.7.1
Limitation on period for starting action	2.13.1
Lump-sum contract	7.1.1

M

Measurement, standard publications on methods of	7.1.5, 7.1.6
Measurement—see also "remeasurement"	
Misrepresentation, effect on contract	1.4(a), 1.4(b)
Mistakes in contract terms	1.4(e)
Modification to contract conditions	2.8.3, 2.8.5, 3.2, 6.1, 6.2

N

Negotiated contract	3.3.1, 7.1.6, 7.2.2, 7.2.5
Negotiated tender	9.3
Nominated sub-contractor	2.6.2, 6.6, 8.3

O

Open tender	9.1
Order, Purchase	5.2, 10
Osborne formula for CPA	7.1.2

P

Package contract	7.2.3
Parties to contract	Def.
Parties to contract, identity of	1.4
Parties to contract, capacity of	1.5, 2.13, 4
Performance Bond	2.11.2
Performance, specific	2.11.2, 2.11.3
Plant performance bond	2.11.2
Plant performance, tests of	2.11.2(d)
Preliminary tender (enquiry)	3.3.1, 9.2
Prequalification Bids	9.5
Price-adjustment contract	7.1.2
Price-adjustment formulae	7.1.2
Prime-cost items	2.6.2, 7.1.5
Programme of contract works	2.7
Provisional sums	2.6.3, 7.1.5
Purchase order, forms of	5.2, 10
Purchase, standard conditions of	6.8

Q
Quantities, Bills of   2.6, 6.3, 7.1.5

R
Rates, Schedule of, contract	2.6.1
Regulations, site	2.9
Remeasurement of quantities	2.6.1, 6.3, 6.6, 7.1.4, 7.1.5
Repayment Bonds	2.11.2
Running contract	7.2.7

S
Sale of Goods Act (1893)	2.8.4
Schedule of rates	2.6.1
Schedule of rates contract	7.1.6, 7.2.3
Scope of contract	2.3
Seal, documents under	2.11.1, 2.13
Security for due performance	2.11.2
Selective Tender	9.2
Serial contract	7.2.6
Serial tender	9.4
Service contract	7.2.8
Signing of contract ("under hand")	2.13
Signing of contract, persons competent to sign	1.5
Site Data	2.4
Site Regulations	2.9
Specification, technical	2.5, 2.6, 2.10, 3.2, 3.3.2, 7.2.3
Specification, technical, modifications to	see "variations"
Stamp, Revenue Duty on Agreements	2.13.3
Standard Forms of Conditions of Contract	2.8, 6, 7.2.4
Sub-contractor, approval of by Engineer	8.2
Sub-contractor, nominated	2.6.2, 6.6, 8.3
Surety for bond	2.11

T
Target-cost contract	3.3.1, 7.1.4, 9.3
Technical specification	2.5, 2.6, 2.10, 3.2, 3.3.2, 7.2.3
Technical specification, modifications to	see "variations"
Tender	2.6.1, 2.10
Tender bond	2.11.2, 2.11.3
Tender date	2.7.1, 2.10.1
Tender deposit	2.11.2
Tender, form of	2.9, 2.10.1
Tender, negotiated	9.3
Tender, open	9.1
Tender, preliminary	3.3.1
Tender, selective	9.2
Tender, serial	9.4
Tender, target cost	3.3.1, 7.1.4
Tender, two-stage	3.3.2

Termination of contract, right of 1.4, 1.9, 2.7.1, 2.11.2
Turnkey Contract 7.2.4

U
Undue influence, effect on contract 1.4(c)

V
Validity period of tender 2.10.1, 2.11.2, 3.4
Variation Clause 2.14.1
Variations of Contract Specification 2.6.1, 2.6.3, 2.14
Variations of Contract terms and conditions 2.14
Verbal contracts 1.8

W
Warranty 1.9, 2.11.1
World Bank, The 3.3.2, 6.11.2, 9.5
Writing, contracts in 1.8
WTBA formulae for CPA 7.1.2